Cambridge astrophysics series

The solar granulation

Other books by the authors

Sunspots, R. J. Bray and R. E. Loughhead (London: Chapman & Hall, 1964; republished by Dover Publications, Inc. 1979)

The Solar Chromosphere, R. J. Bray & R. E. Loughhead (London: Chapman & Hall, 1974)

Illustrated Glossary for Solar and Solar-Terrestrial Physics, A. Bruzek & C. J. Durrant (eds.) (Dordrecht: D. Reidel Publishing Co. 1977)

The solar granulation

R. J. BRAY, R. E. LOUGHHEAD
AND C. J. DURRANT

SECOND EDITION

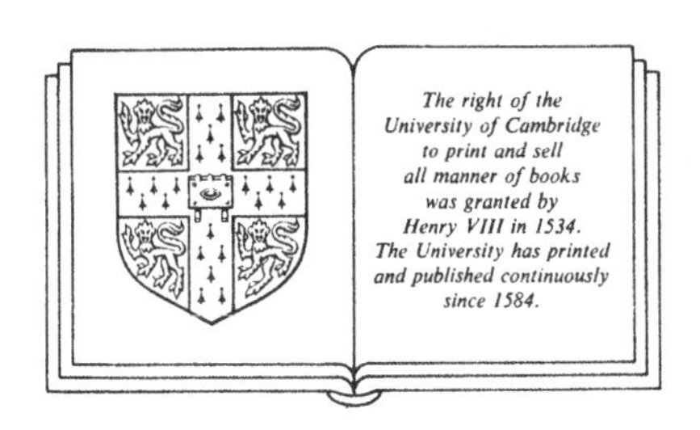

CAMBRIDGE UNIVERSITY PRESS
Cambridge
London New York New Rochelle
Melbourne Sydney

CAMBRIDGE UNIVERSITY PRESS
Cambridge, New York, Melbourne, Madrid, Cape Town, Singapore, São Paulo, Delhi

Cambridge University Press
The Edinburgh Building, Cambridge CB2 8RU, UK

Published in the United States of America by Cambridge University Press, New York

www.cambridge.org
Information on this title: www.cambridge.org/9780521115513

First published by Chapman & Hall 1967
Second edition published by Cambridge University Press 1984
This digitally printed version 2009

A catalogue record for this publication is available from the British Library

Library of Congress Catalogue Card Number: 83-1881

ISBN 978-0-521-24714-6 hardback
ISBN 978-0-521-11551-3 paperback

CONTENTS

PREFACE

When examined with a sufficiently powerful telescope under conditions of good atmospheric seeing the surface of the Sun reveals a fine structure consisting of an irregular cellular pattern of polygonal bright elements - granules - separated by narrow dark lanes. Over the whole surface there are some four million granules, whose average diameter is only about $1''.3$–$1''.4$ of arc (~ 1000 km). The observation of such small structures is not easy, and only in the last twenty-five years or so has the application of high-resolution techniques (both from the ground and from high-altitude balloons) given us a detailed picture of the properties and mode of origin of the granulation. Since the publication of the first edition of this book in 1967, the pace of advance has quickened: granule observations of higher resolution, covering longer periods of time, have been obtained and powerful computers are opening the way to realistic numerical simulations of the complex hydrodynamic processes involved. Our aim in this revised and expanded edition is to present a comprehensive and updated account of existing observational and theoretical knowledge.

Serious interest in the fine structure of the solar surface dates back to the beginning of the nineteenth century, and in a brief historical introduction (Chapter 1) we trace the changing ideas concerning the nature of the granulation resulting first from visual examination and, later, from photographic and spectroscopic observations, during the period 1801–1957. The latter date marks the beginning of the modern era, and in Chapter 2 we give a detailed account of the knowledge derived since then of the morphology, evolution, and dynamics of the photospheric granulation. In addition, we digress slightly in order to consider the corresponding properties of the supergranulation, which bears a certain resemblance to the ordinary granulation and may have a similar mode of origin.

Up to the present, our knowledge of the granulation has been based solely on observations from either ground-based or balloon-borne telescopes. However, many of the problems still confronting us require observations lying at or beyond the limit of the technically feasible. It is true that towards the late 1980s we can look forward to higher-resolution instruments carried on satellites, e.g. NASA's 1.25-m Solar Optical Telescope (SOT). Nevertheless, for quite some time to come we shall continue to rely very heavily on observations from the ground. Thermal disturbances in the Earth's atmosphere impose fundamental limitations on such observations and therefore we begin Chapter 2 with a brief review of solar seeing and its implications for solar observing.

Modern observations show unmistakably that the photospheric granulation is basically a convective phenomenon, each granule and its surrounding dark material representing a single convection cell. In fact, the granules are the visible manifestation of subphotospheric convection currents which contribute substantially to the outward transport of energy from deeper layers and thus help to maintain the energy balance of the Sun as a whole. Actually, they play an even wider role since, in the upper levels of the convection zone, they are believed to give rise to waves which are partially, if not wholly, responsible for heating the overlying chromosphere and corona. A discussion of the various possible types of waves lies beyond the scope of this book, but a detailed account can be found in a companion monograph (*The Solar Chromosphere*, R. J. Bray & R. E. Loughhead, London: Chapman & Hall).

A knowledge of the modern theory of fluid convection is evidently an essential pre-requisite to an understanding of convection in the Sun and stars and therefore to a proper interpretation of the solar granulation. Accordingly, in Chapter 3 we give an introduction to this theory, laying emphasis on those aspects of particular relevance to astrophysical systems. The topics covered include the basic hydrodynamic equations, the anelastic and Boussinesq approximations, the classical Rayleigh problem and its extension to stratified fluids, and laboratory convection. Although most laboratory experiments have been restricted to incompressible fluids, they can nevertheless provide valuable insight into the onset of convection in *compressible* fluids and into the processes involved in the transition to developed and ultimately, at very high Rayleigh numbers, turbulent convection. The chapter ends with a discussion of the so-called 'ice-water' experiment which, paradoxically, provides illuminating insight into the overshooting of the convective motions at the top of the solar convection zone into the stably stratified layer above.

In Chapter 4 we turn from laboratory to astrophysical convection and consider the theory of developed convection in a *stratified* gas. We start by describing different versions of the well-known mixing-length theory, which has been widely used to provide zeroth-order models of stellar convection zones. Next we consider more rigorous theories, seeking 'exact' solutions to the non-linear hydrodynamic equations which describe the flow pattern in detail. In so doing we have to recognize that solar convection is turbulent in character and therefore an understanding of the interplay between the large-scale ordered motions and the small-scale turbulence is vital. As one rises through the granulation layer the solar gas changes from being optically thick to optically thin and radiative smoothing of temperature differences between different parts of the gas becomes increasingly effective. Accordingly, our treatment of the theory of astrophysical convection ends with a discussion of radiative energy transfer in such an inhomogeneous medium.

The subject matter of Chapters 3 and 4 is inherently difficult and complex, but we have tried to develop the account in such a way as to make the subject accessible to solar physicists and astrophysicists with no expert knowledge of modern fluid mechanics and to theoretical hydrodynamicists with little or no previous contact with solar physics.

It must be admitted that the modern theory of astrophysical convection still suffers from certain deficiencies, e.g. the absence of a comprehensive treatment of the non-linear interactions leading to stochastic behaviour in the small-scale motions. Nevertheless, its application to the Sun has significantly advanced our understanding of the physical nature of the solar granulation and the solar convection zone. In the final chapter, Chapter 5, we confront the theory with the results of the extensive observations described in Chapter 2 wherever this is both possible and fruitful. The topics discussed include the mean cell size, lifetime, granulation near the extreme limb, height of overshoot, convective heat flux, direction of cellular motion and turbulence spectrum.

The comparison between observation and theory (Chapter 5) is effected by the construction of inhomogeneous models, taking account of the horizontal and vertical variations in the various hydrodynamic and thermodynamic quantities. This approach has advanced to the stage where, with the aid of modern large computers, a start has been made on the time-dependent numerical simulation of the granulation phenomenon. With the coming generation of computers such simulations should become increasingly realistic. But already a consistent picture of the granulation has emerged. This is important, not least because a convincing numerical simulation of the solar convection zone would constitute almost the sole astrophysical test of the basic theory of convection in highly stratified fluids.

The Sun is in no sense a peculiar star: a dwarf of spectral class G2 it is – as far as we know – a typical member of the lower main sequence of the Hertzsprung-Russell diagram. If we were able to examine the surfaces of other stars of similar spectral class, we would presumably see the phenomena of the granulation and supergranulation in much the same form. In fact, stars covering a wide range of spectral class are believed to possess convective envelopes similar to that of the

Sun. G, K and M stars have a hydrogen convection zone of considerable thickness, while A stars have a thin zone at the base of the atmosphere, the transition from a thick to a thin zone occurring at class F. The early B stars possess, instead of a hydrogen convection zone, a weak thin helium convection zone which carries only negligible heat. Among red stars, both giants and dwarfs have a deep convection zone, which in the extreme case of a late M dwarf extends all the way to the star's centre. In general, a convective core is associated with a strong concentration of the thermonuclear energy generation toward the centre, which is typically the case in early-type stars; the Sun is believed not to have a convective core.

On the other hand the Sun is the only star on which we are ever likely to see resolved granules. Therefore, information about stellar granulation can come only from Fraunhofer line profiles integrated over the stellar disk. Two promising diagnostics for the presence of stellar convection are discussed at the end of Chapter 5, based on our knowledge of the asymmetries and shifts of solar lines. Although the effects are small and refined measuring techniques are required, attempts to exploit them are in progress. In fact, we are witnessing the birth of a new era in stellar physics: more rigorous theories of astrophysical convection are being applied to the calculation of stellar models and, for the first time, the effects of convection in stellar atmospheres are being observed and compared with theory.

R.J.B. R.E.L. C.J.D.

Note by C. J. Durrant

This edition was conceived in mid-1979 when I was invited by the authors of the first (1967) edition to prepare a revision which would involve updating the observational parameters, reworking the discussion of convective theory and excising material which appeared dated or had been superseded. None of us then appreciated quite how much had been published in the intervening years, not only in terms of sheer volume but also in range of content. As a result, the framework of the book had to be enlarged in order to accommodate a treatment that would do justice to recent developments. Thus the reader will find two chapters devoted to convective theory in place of one previously; these are balanced by a chapter on the observed properties of the granulation which has been considerably enlarged to take account of its statistical as well as individual characteristics. The theoretical and observational material is drawn together in the final chapter to provide as complete a physical description of the granulation phenomenon as is possible today.

The bulk of this new material has been drafted by myself but the final text represents the outcome of an extended reworking by all three authors. To end on a personal note, I would like to express my warmest thanks to my coauthors. Not only did they encourage me to proceed with a more radical revision than was originally envisaged, but they were a constant source of material, advice, criticism and tact. The collaboration was most stimulating and rewarding; I hope that the reader will discern in the text some traces of the pleasure that the preparation of this edition has provided.

C.J.D.

ACKNOWLEDGEMENTS

The authors wish to thank Mrs L. Schienagel and Mrs I. David for their skilful preparation of many of the figures. We are grateful to Mrs A. C. Durrant for her careful typing of the manuscript and indefatigable editorial assistance, not least in obtaining and checking references and preparing the indices.

For supplying illustrative material we thank the following persons and organizations:

Dr G. Ceppatelli, Arcetri Observatory (Fig. 2.14).
Dr V. N. Karpinsky, Pulkovo Observatory (Figs. 2.2, 2.16(*b*)).
Professor R. B. Leighton, California Institute of Technology (Fig. 2.26).
Dr W. C. Livingston, Kitt Peak National Observatory (Fig. 2.16(*a*)).
Professor W. Mattig, Kiepenheuer Institut (Fig. 2.4(*b*)).
Dr J. P. Mehltretter, Kiepenheuer Institut (Fig. 2.12).
Professor M. Schwarzschild, Princeton University (Fig. 2.1).
Dr H. Wöhl, Kiepenheuer Institut (Fig. 2.4(*a*)).
Sacramento Peak Observatory (Fig. 2.25).

For permission to reproduce copyright material we are indebted to Cambridge University Press, D. Reidel Publishing Co. and Springer-Verlag, and to the editors of the following journals: *Astronomy and Astrophysics*, *Astrophysical Journal* and *Zeitschrift für Astrophysik*. For permission to make use of other illustrative material we thank P. N. Brandt, E. Graham, D. F. Gray and J. P. Mehltretter.

Our sincere thanks go to a number of friends and colleagues in Australia, Germany and the USA (in addition to individuals specifically acknowledged in the text) who, by private discussions and correspondence, have helped to clarify our ideas concerning some of the difficult observational and theoretical problems dealt with in this book. We thank especially those members of the Kiepenheuer Institut who provided so much stimulus and criticism.

Finally, we express our gratitude to the chief of the CSIRO Division of Applied Physics, Dr J. J. Lowke, and to the director of the Kiepenheuer Institut für Sonnenphysik, Professor E. H. Schröter, for support.

Commonwealth Scientific and Industrial Research Organization, National Measurement Laboratory, Sydney, Australia — R. J. Bray, R. E. Loughhead

Kiepenheuer Institut für Sonnenphysik, Freiburg, West Germany — C. J. Durrant

1

HISTORICAL INTRODUCTION

1.1 Early visual observations of the photospheric granulation

No serious attempt to elucidate the fine structure of the solar surface was made until the beginning of the nineteenth century, when the problem attracted the attention of the famous English astronomer Sir William Herschel (1738–1822). Observing the Sun with a reflector of focal length about 3 m fitted with a speculum mirror of his own manufacture, Herschel interpreted what he saw in the light of his own highly exotic views regarding the habitability of the Sun. 'On a former occasion', he wrote in 1801, 'I have shewn that we have great reason to look upon the Sun as a most magnificent habitable globe'. Herschel pictured the solar disk as being covered by *corrugations* which, he said, 'I call that very particular and remarkable unevenness, ruggedness, or asperity, which is peculiar to the luminous solar clouds, and extends all over the surface of the globe of the Sun. As the depressed parts of the corrugations are less luminous than the elevated ones, the disk of the Sun has an appearance which may be called mottled'. From this description it seems clear that Herschel did not resolve the individual photospheric granules as such but rather the large-scale pattern of brightness fluctuations which appears when the granulation is viewed with inadequate resolving power or under mediocre conditions of atmospheric seeing.

After Herschel, interest in the problem of the fine structure of the solar disk languished until, in the early 1860s, it suddenly became the centre of a spirited controversy involving many of the foremost solar observers of the day. The originator of the controversy was the English engineer James Nasmyth (1808–90), who is remembered as the inventor of the steam hammer and as an assiduous observer of the Moon. At his factory at Bridgewater, near Manchester, Nasmyth had all the facilities for casting and polishing specula. With his largest mirror, a 50-cm, he constructed a Cassegrain–Newtonian telescope on an alt-azimuth mounting. After his retirement to Penshurst in Kent, Nasmyth used this telescope to make many observations of sunspots and the solar photosphere, using a high-powered eyepiece and choosing moments when the seeing was best.

These observations led him to announce in 1862 that the Sun's surface was actually covered by a compact pattern of thin bright filaments shaped much like 'willow-leaves'. According to Nasmyth, the 'willow-leaves' were extremely regular in shape and size but crossed one another in all possible directions, the dark interstices between them giving rise to the mottled appearance of the disk. Nasmyth's conception of the willow-leaf pattern is illustrated in Fig. 1.1, which is a reproduction of a drawing that he made on 5 June 1864 of a sunspot group and the surrounding photosphere.

Nasmyth's announcement of his discovery of the willow-leaf pattern sparked off a number of searching discussions throughout the astronomical world. One experienced English observer, the Reverend William Dawes (1799–1868), flatly denied the existence of such structures; he stoutly maintained the view (which he considered well established) that the mottling of the solar surface shows every variety of irregular form except in the immediate vicinity of sunspots. The observers at the Royal Greenwich Observatory claimed to corroborate Nasmyth, although their description of the pattern in terms of interlaced *rice grains* represented a retreat from Nasmyth's extreme view of the regularity of the basic structure. The First Assistant, E. J. Stone, made grain counts with an altazimuth transit instrument of 9.5-cm aperture that yielded a rough dimension of between 1''.5 and 2'' of arc.

The famous Italian astronomer Father Secchi (1818–78) derived from his own observations a picture of the photospheric fine structure which was much closer to reality. He described the solar surface as being covered by a multitude of small bright features, which he also likened to grains, separated by lanes of darker material. The grains were similar in size but differed considerably in shape. Fig. 1.2 is a reproduction of one of Secchi's own drawings taken from his treatise *Le Soleil*, showing the pattern in the neighbourhood of a small sunspot pore.

While the morphological details of Secchi's drawing constitute a fair approximation to the truth, the same cannot be said of his attempt to measure the sizes of the individual grains. He derived a figure of about 0''.3, whereas modern observations yield a representative value for the diameter of the granules of 1''.3 (Section 2.3.2). Similar measurements were made by another assiduous visual observer, S. P. Langley (1834–1906), who in 1867 was appointed director of the Allegheny Observatory in the USA and later became secretary of the Smithsonian Institution and a notable pioneer in the new science of aerodynamics. Using the 33-cm Allegheny refractor, he found that the average diameter of the grains was between 1'' and 2'' but claimed that, at the moments of best seeing, the individual grains appeared as conglomerates of smaller elements not exceeding 0''.3 or 0''.4 in width.

Fig. 1.1. Drawing of a large sunspot group and the surrounding photosphere made by James Nasmyth on 5 June 1864, illustrating his conception of the appearance of the solar disk as a concentrated pattern of thin bright filaments or 'willow-leaves'. According to Nasmyth, the 'willow-leaves' were extremely regular in shape and size but crossed one another in all possible directions, the dark interstices between them giving rise to the mottled appearance of the disk.

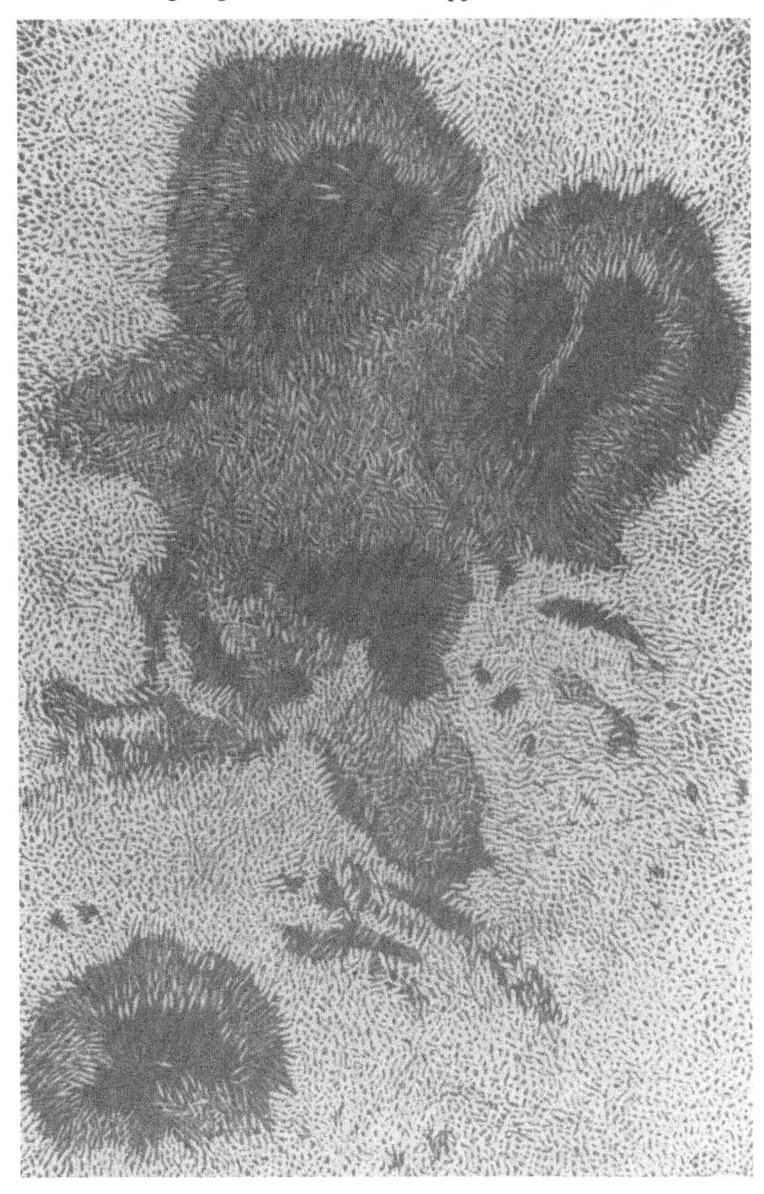

The visual observer whose description of the photospheric fine structure came closest to the truth was the English astronomer Sir William Huggins (1824–1910), who is remembered primarily for his important contributions to the field of stellar spectroscopy. According to Huggins the grains, which he preferred to call

Fig. 1.2. Drawing of the photosphere in the neighbourhood of a sunspot pore published by Father Secchi in his book *Le Soleil* (1875). He pictured the solar surface as covered with a multitude of small bright features ('grains') separated by lanes of darker material. Secchi's grain structure bears a fair resemblance to the photospheric granulation pattern although, unfortunately, no scale was given on the original drawing.

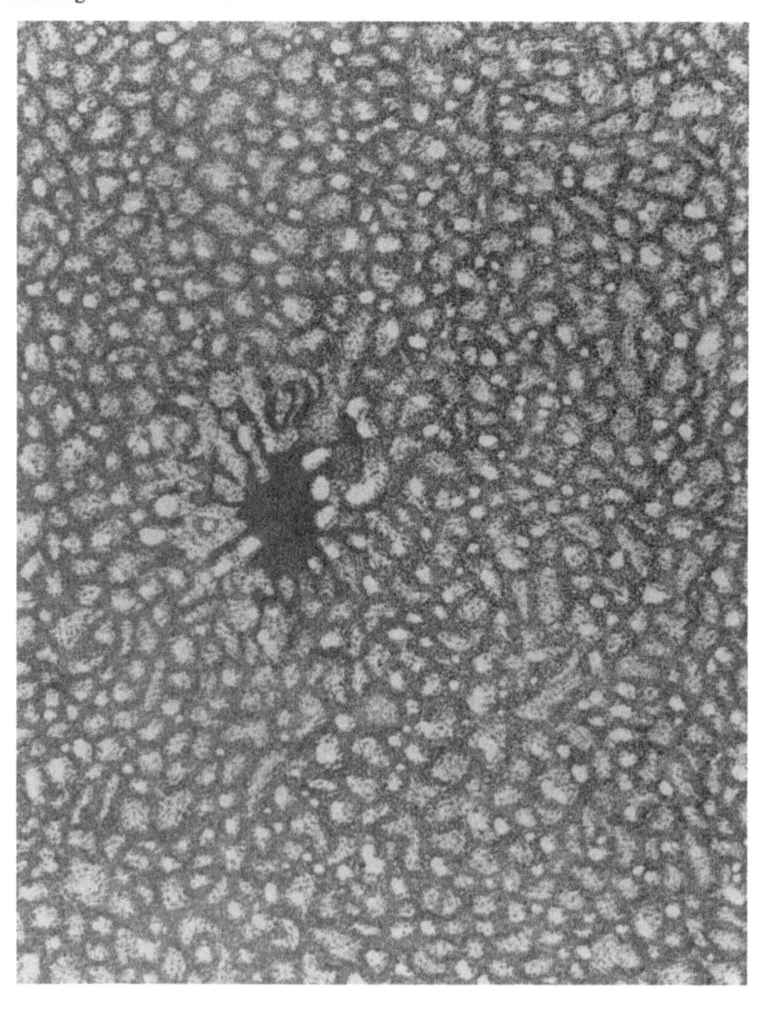

by the name *granules* suggested by Dawes in 1864, were distributed over the entire solar surface and were more or less round or oval in shape, although more irregular forms did occur. Their diameter he estimated to lie between 1″ and 1″.5, in agreement with the modern figure. Huggins' note published in 1866 in the *Monthly Notices of the Royal Astronomical Society* effectively terminated the controversy over Nasmyth's 'willow-leaves'. Unfortunately, however, Huggins went beyond his realistic description of the individual granules to claim that they were grouped to form a variety of fantastic shapes and patterns; these he illustrated by a drawing, which the interested reader will find reproduced in Young's book, *The Sun*.

1.2 Pioneering photographic observations of Janssen, Hansky, and Chevalier

On 13 August 1877 the French astronomer Pierre Jules Janssen (1824-1907) rose before a meeting of the Academy of Sciences in Paris and announced that he had successfully photographed the photospheric granulation. Born the son of an eminent musician and educated at the University of Paris, Janssen achieved fame by the discovery in 1868 - made independently and nearly simultaneously by Sir Norman Lockyer - that with the aid of a spectroscope it was possible to see prominences outside of an eclipse. As a result of this work he was appointed director of a new astrophysical observatory set up at Meudon, in the vicinity of Paris (Fig. 1.3). There, using a refractor of 13.5-cm aperture, he quickly brought the art of high-resolution solar photography to a high degree of perfection. In fact, when the American astronomer Langley visited Meudon in 1877, he remarked that during his many years of visual observations there had been only five or six occasions when he had seen photospheric detail with a clarity equal to that of Janssen's photographs - and then only for a few seconds at a time.

Fig. 1.4 is a reproduction of part of a photograph obtained by Janssen on 1 April 1894. It shows clearly that the photospheric granulation in the central region of the solar disk consists of a well-defined pattern of bright granules with diameters lying mostly in the range 1-2″, separated by lanes of darker material. The resolution is, in fact, comparable to that of modern granulation photographs taken with telescopes of similar aperture. A number of factors contributed to Janssen's success: the choice of a telescope of a type suitable for high-resolution solar photography; the use of an enlarging lens to obtain a large effective image diameter (30 cm) with a relatively short optical path; and the employment of a 'flying-slit' shutter to achieve very short exposure times. Yet despite these notable advances in technique, Janssen's observations contributed very little to our knowledge of the properties of the individual photospheric granules, principally because he was side-tracked into devoting most of his effort to studying a large-scale pattern of distortions occurring on many of his photographs, to which he gave the name *réseau photosphérique*. Janssen believed this pattern to be an actual feature of the solar surface, produced by violent movements of the granules in certain localized areas. However, the work of many subsequent observers has conclusively shown that the *réseaux* are due entirely to the effects of poor atmospheric seeing.

Fig. 1.3. Pierre Jules Janssen, the famous French astronomer, who was the first to photograph successfully the photospheric granulation. His statue stands in the grounds of the Meudon Observatory, with the city of Paris in the background.

In all, Janssen's photospheric observations extended over a period of some twenty years. They were published in collected form in 1896 in a volume which also contains reproductions of twelve of his original photographs. Plate X of this collection is well known: it is the one on which the granulation appears to show a striking resemblance to a very regular polygonal convection pattern. (Part of

Fig. 1.4. Reproduction of part of a photograph of the photospheric granulation obtained by Janssen on 1 April 1894. Although marred by seeing, it shows clearly that the granulation consists of a well-defined pattern of bright granules mostly 1–2″ of arc in diameter, separated by lanes of darker material.

this photograph has been reproduced by Kiepenheuer, 1953: see Fig. 13.) However, in the light of modern knowledge, we must regard this particular photograph with a good deal of suspicion. In the first place, the granulation pattern appears to be *too* regular, whereas we now know that there is actually a considerable diversity in the sizes and shapes of the individual granules (cf. Figs. 2.1 and 2.2). Secondly, a careful examination of Janssen's original reproduction shows that in the regions of the photograph where the regular polygonal appearance is most marked, the individual features have diameters lying mostly in the range 2–4″, according to the scale given by Janssen. These features are therefore about twice the size of normal photospheric granules and we must consequently attribute their origin to some spurious effect. One possible explanation may lie in the action of surface tension forces during the processing of the original plate, which was of the wet collodion type. Drying paint films, for example, often display regular, convection-like patterns produced by variations in surface tension (see Section 3.4.1).

The first observer to point out the atmospheric origin of Janssen's *réseau photosphérique* was the young Russian astronomer Alexis Hansky (1870–1908). Educated at the University of Odessa, he spent some time at the Meudon Observatory before returning home to Russia where, in 1905, he was appointed an assistant astronomer at the Pulkovo Observatory near Leningrad. There, using a conventional astrograph and an enlarging camera, he turned his attention to high-resolution photography of the solar disk. His main interest lay in the photospheric granulation, and he succeeded in obtaining granulation photographs showing somewhat better resolution than those of Janssen himself. Moreover, Hansky deserves special credit for realizing the importance of obtaining sequences of granulation photographs taken at short time intervals apart in order to study changes in the granules with time. Although his attempts to obtain sequences were only partially successful, Hansky nevertheless derived an estimate of about 5 min for the mean lifetime of the photospheric granules - about one-third of the correct value (cf. Section 2.3.6). On the other hand, he erroneously concluded that the granules execute horizontal oscillations about a mean position with speeds of up to 4 km s^{-1}, failing to appreciate that their apparent lateral displacements from photograph to photograph were due - like Janssen's *réseaux* - entirely to the effects of atmospheric seeing (see Section 2.2.1).

In 1908 Hansky went to the Crimea, where the Pulkovo Observatory had recently established a southern station, with the intention of continuing his observations of the granulation under more favourable climatic conditions. However, just as his programme was about to start, he tragically lost his life while bathing in the Black Sea.

A third pioneer in the art of high-resolution solar photography was Father Stanislas Chevalier (1852–1930), for many years the director of the Zô-Sè

Observatory in China. This observatory was founded as a branch of an older Jesuit observatory situated at Zi-Ka-Wei on the outskirts of Shanghai. The Zô-Sè Observatory was built on a low hill some 24 km from Shanghai and was equipped with twin 40-cm refractors of 7-m focal length carried on the same equatorial mounting, one designed for visual work and the other for photography. Although the instrument was brought into operation in 1901, observations of the Sun did not start until 1904. Chevalier's photographs, like those of Janssen and Hansky, clearly demonstrated that the photospheric granulation consists of a pattern of bright granules mostly 1–2″ in diameter, separated by lanes of darker material. However, although Chevalier devoted much effort to the study of the granulation, his work added little to the existing knowledge of its properties. In fact, Chevalier's claim to recognition rests more on his contributions to the study of the fine structure of sunspots than on his work on the granulation.

As a result of the pioneering photographic observations of Janssen, Hansky, and Chevalier, by 1914 the existence and nature of the photospheric granulation was firmly established. However, the time was not yet ripe for a proper physical interpretation of the phenomenon. For example, Chevalier was content to conclude a paper on the photosphere published in *The Astrophysical Journal* in 1908 in the following vein:

> Let us admit ... that the granules are the summits of a fleecy stratum of condensed particles, with or without any horizontal movement; and that the stratum is subject to undulatory movements; the summits of the waves will then present the same succession of changes, their relative position varying in every direction and with any velocity. The short, quickly changing waves of a choppy sea may possibly give us a faint imitation of what is realized on a gigantic scale and in a very different element in the solar photosphere.

Despite their achievement in removing any doubt possibly remaining as to the existence and form of the granulation pattern, Janssen, Hansky, and Chevalier failed to exploit the full potentialities of high-resolution photography. Moreover, they did not realize just how much physical information could be gained from observations of this kind. In fact, nearly half a century was to elapse before the new technique, in an improved form, was fully utilized as a research tool for the study of the solar photosphere (see Section 1.6).

1.3 Strebel's discovery of the polygonal nature of the granules

Following the pioneering work of Janssen, Hansky, and Chevalier, interest in high-resolution photography of the photospheric granulation waned and it was not until 1933 that an important new observational discovery was announced. This came as a result of the efforts of a German physician and amateur astronomer, Hermann Strebel (1868–1943), who, in collaboration with a technician, B. Schmidt, photographed the granulation with a 35-cm horizontal reflecting telescope belonging to the Munich Observatory, diaphragmed down to 20 cm. Amongst other things, Strebel paid particular attention to the important question of the true shape of the granules and reached the conclusion, to quote his own words, '*daß tatsächlich die Granula der Hauptsache nach polygonale Gebilde sind*, daß selbst ausgesprochene Dreiecksquerschnitte häufig vorkommen' (in translation: '*that the granules are actually in the main polygonal structures*, and even definitely triangular structures frequently occur'). The polygonal outlines of the granules are clearly evident on some of the photographs published by Strebel, the best of which show areas where the resolution is, in fact, comparable to that of good modern photographs (see Section 2.3.1).

Strebel's observations demonstrated more clearly than ever before the striking resemblance of the photospheric granulation to an irregular, cellular *convection pattern*. However, Strebel's discovery, despite its importance and its publication in a well-known international journal (*Zeitschrift für Astrophysik*), apparently attracted little attention and, as the years passed, was largely forgotten. In fact, as we shall see in Section 1.6, the irregular, polygonal character of the granulation pattern was not re-discovered until 1957, when photographs of the photosphere were for the first time obtained with a resolution surpassing that achieved by Schmidt and Strebel a quarter of a century before.

1.4 Identification of the granules as convection cells

The essential foundation for the modern convective theory of the origin of the photospheric granulation was laid by the distinguished German astrophysicist Albrecht Unsöld, who in 1930 showed that, as a consequence of the increase in hydrogen ionization with depth, there must exist a zone of convective instability directly beneath the visible photospheric layers. As we shall explain in more detail in Section 5.2.1, an elementary volume of gas moving upwards through the hydrogen ionization zone is heated by the release of ionization energy. The buoyancy of the element is thus increased and it continues its upward journey. In this way convection currents are generated, to which Unsöld attributed the origin of the photospheric granulation (and, incidentally, of sunspots as well).

Unsöld himself did not attempt to give any detailed picture of the exact mechanism involved, but this task was soon taken up by other workers. In 1933 H. Siedentopf (1906–63) suggested that the granules represent globules or bubbles of hot gas pushing their way upwards through cooler descending material. He pointed out that the value of the Reynolds number under the conditions obtaining in the hydrogen convection zone exceeded the critical figure for the onset of turbulence and concluded that the convection currents must therefore be turbulent in character. (The Reynolds number criterion for the onset of turbulence is an empirical rule valid in the case of the forced flow of an incompressible fluid through a pipe; criteria more relevant to free convection involve the Rayleigh and Prandtl numbers introduced in Chapter 3.)

Siedentopf based his ideas on the 'mixing-length' theory of turbulence introduced shortly before by the famous German aerodynamicist L. Prandtl. The fundamental basis of this theory is the assumption that the convective energy is carried by 'turbulent eddies' which part with their energy and momentum after travelling a distance equal to the mixing-length. Siedentopf identified the granules with Prandtl's turbulent eddies, and his ideas were refined and elaborated by a number of later workers, including R. v. d. R. Woolley, L. Biermann, C. de Jager, and E. Böhm-Vitense.

Today, the mixing-length or bubble theory of convection is no longer regarded as valid, since it ignores virtually all details of the hydrodynamic processes that occur in a convecting fluid. However, it survives as a relatively simple method of calculating zeroth-order approximations to models of the convection zones of the Sun and stars, although in recent years even this application has been subjected to an increasing amount of criticism (see Section 4.2).

A very different picture of the type of convective motion responsible for the granulation was put forward in 1936 by the Oxford astronomer H. H. Plaskett. He drew attention to the classic laboratory experiments on convection in thin liquid films performed by H. Bénard at the beginning of the century and pointed out the striking resemblance between the appearance of the granulation and the cellular convection patterns observed in liquids heated from below. This resemblance led Plaskett to suggest that each granule and the surrounding intergranular region be identified with a Bénard cell in the unstable hydrogen ionization zone. The idea of identifying the granules with convection cells was, in fact, a direct forerunner of present-day theoretical research into the nature of the convective processes responsible for the granulation, a topic which forms the subject matter of Chapters 3 and 4. However, owing perhaps to the formidable mathematical difficulties associated with the hydrodynamic theory of cellular convection, Plaskett's suggestion was at the time largely ignored by theoreticians who, for more than two decades, continued to concentrate their attention on the mixing-length theory. The one noteworthy exception was J. Wasiutynski who, in a long and rather unorthodox treatise published in 1946, severely criticized the mixing-length theory and championed the Bénard cell interpretation.

Today we are in no doubt that the granules are to be identified as convection cells, but we know that the type of convection involved is not of the steady Bénard variety.

1.5 First spectroscopic measurements of granule velocities; attempts to interpret the granules as 'turbulent eddies'

In the years immediately following the end of the Second World War considerable advances were made in the production of plane diffraction gratings of large size and high light-efficiency. This development was pioneered at the Mt Wilson Observatory by H. D. Babcock and was subsequently carried on by his son, H. W. Babcock, from 1963 to 1978 director of the Mt Wilson and Palomar Observatories. Besides having great spectroscopic resolving power, the gratings made by the Babcocks were blazed to diffract a large percentage of the incident light into a given order and so enabled high-dispersion solar spectra to be taken with much shorter exposure times than was previously possible. With the shorter exposures the degrading effect of atmospheric seeing on the spatial resolution of the spectra was reduced, and in this way it became possible to obtain spectra with a definition approaching that shown by good-quality direct photographs of the photosphere - albeit only under the best seeing conditions.

One of the first large solar instruments to benefit from the advent of the blazed diffraction grating was the 23-m spectrograph of the 46-m solar tower at Mt Wilson. In 1949 a new grating, ruled under the supervision of H. D. Babcock, was installed in the spectrograph and with it R. S. Richardson succeeded in obtaining for the first time spectra having a spatial resolution sufficient to resolve individual photospheric granules. These spectra showed prominent bright and dark streaks running parallel to the dispersion, which were produced by the granules and intergranular dark spaces falling on the spectrograph slit. The streaks themselves were intersected by the solar absorption lines in a definite zigzag manner, indicating the presence of both upward and downward velocities at the points of intersection (see Fig. 2.16). The magnitude of the velocities was of the order of a few tenths of a kilometre per second.

Working in collaboration with the theoretical astrophysicist Martin Schwarzschild of the Princeton University Observatory, Richardson used one spectrum showing exceptionally fine definition to compare the Doppler displacements along the length of the slit with the brightness fluctuations at the corresponding points in the continuum. *A priori*, on the basis of a convective explanation of the origin of the granulation, one would have expected a strong correlation between the brightness and velocity variations, the hot granules moving upwards and the cool intergranular material downwards. On the contrary, however, Richardson and Schwarzschild found that the observed correlation was rather weak, only the narrow regions of high upward velocity appearing to be systematically brighter than average.

At the time this result seemed to throw some doubt on the identification of the granules as convection cells. Since then, however, the problem of granule velocities has been attacked by a large number of other workers using more powerful observing techniques and, as a result, it is now known that the granule velocities Richardson and Schwarzschild attempted to measure are masked by an oscillatory velocity field that is independent of the granular field. When the spectroscopic measurements are interpreted in the light of this knowledge, they do not actually throw doubt on the convective origin of the granulation. Nevertheless, the questions of the true nature of the small-scale velocity field observed in solar absorption lines and of its relationship to the granulation have proved to be of considerable complexity; both topics are discussed in Sections 2.4.2, 2.5.4, and 2.5.5.

One fact which greatly puzzled Richardson and Schwarzschild was the smallness of the measured granule velocities compared with estimates previously obtained by a variety of indirect means, such as studies of the photospheric curve of growth and line profile analyses. This consideration led them to advance the hypothesis that the motions in the photosphere, far from being basically convective, were in fact a manifestation of a *large-scale turbulence*. On this basis they attributed the discrepancy in the velocity estimates to the fact that the various determinations referred to 'turbulent elements' of different sizes. Their argument is perhaps best summarized in their own words:

> Laboratory experiments have shown that the state of turbulence of a gas cannot be described well by one mean size of the turbulent elements but rather has to be described by a whole continuous spectrum of sizes. In every case of turbulence there appears to exist one size of elements which possesses the highest average velocity. Elements with sizes larger than those of the fastest elements show a rapid decrease of average velocity with element size. On the other hand, toward smaller element sizes the average velocity falls off slowly. More particularly, under conditions usually fulfilled in astronomical cases this fall-off is governed purely by the process of dissipation of bigger elements into smaller elements. In this process, according to Kolmogoroff, the random turbulent velocity decreases as the reciprocal cube root of the element size.
>
> Richardson & Schwarzschild (1950) *Astrophysical Journal*, **111**, 357.

Using Kolmogoroff's law, Richardson and Schwarzschild combined the results of the various velocity determinations mentioned above to derive a 'spectrum of turbulence' for the solar photosphere. The form of the turbulent velocity spectrum suggested that the diameter of the turbulent elements of highest average velocity should be about 100 to 200 km – less than one-fifth of the diameter of the photospheric granules visible on good-quality direct photographs. Richardson and Schwarzschild therefore concluded that their measured values of the granule velocities probably represented merely the statistical effect of the higher velocities of small unresolved turbulent elements, whose existence had hitherto been unsuspected.

Richardson and Schwarzschild also pointed out that these small, energetic elements or granules should have a greater brightness than those hitherto observed and were thus led to predict the existence of granules very much smaller and brighter than any previously resolved on direct photographs. This prediction attracted considerable attention and, in fact, it was in order to test it that, in 1955, Schwarzschild decided to build a 30-cm balloon-borne telescope to photograph the Sun at a height well above the disturbed layers of the atmosphere responsible for poor seeing (see Section 1.6).

The modern reader may feel that Richardson and Schwarzschild erected rather an elaborate superstructure upon somewhat flimsy foundations. However, in order to place their work in its proper historical perspective, one has to appreciate the remarkable extent to which the thinking of many astrophysicists in the years following the end of the Second World War was influenced by the considerable developments then taking place in the theory of aerodynamic turbulence. The advance which attracted the attention of astronomers most was A. N. Kolmogoroff's formulation of similarity laws to describe the statistical equilibrium of the small-scale components of turbulence. (Kolmogoroff's work was published in the USSR in 1941 but did not become generally known outside that country until after the end of the war.) Under appropriate circumstances these laws enable specific predictions to be made about observable properties of turbulent velocity fields even in the absence of a proper understanding of their exact mode of origin. Despite the fact that Kolmogoroff's laws are strictly applicable only to an incompressible fluid in a state of fully developed, homogeneous, isotropic turbulence, there was a rush of indiscriminate attempts to apply them over the whole gamut of astrophysical problems. These ranged from Richardson and Schwarzschild's attempt to construct a complete turbulent velocity spectrum for the solar photosphere to C. F. von Weizsäcker's suggestion that the galaxies or clusters of galaxies represent the largest eddies of a primordial cosmic turbulence!

There is another aspect of Richardson and Schwarzschild's work which, viewed with historical hindsight, seems even more remarkable than the prevailing pre-disposition to the uncritical acceptance of theories based on turbulence. This is the fact that these workers chose to ignore the overwhelming evidence against the existence of large-scale turbulence in the photosphere provided by the results of more than seventy years of granulation photography. These included not only the observations of Janssen, Hansky, Chevalier, and Strebel, already described, but also good-quality photographs obtained by a number of more modern observers, including P. C. Keenan at the Yerkes Observatory, and P. ten Bruggencate, H. von Klüber, and others at the Potsdam Astrophysical Observatory (see the first edition of this book, Table 2.7). The very existence of a distinct cellular pattern of bright granules on a dark background revealed by these observations was, *ipso facto*, conclusive proof that any motions associated with the granulation were basically well ordered and certainly not predominantly turbulent in character. Richardson and Schwarzschild's failure to recognize this fact was tantamount to questioning the validity of the observational evidence for the existence of the granulation pattern as such. Indeed, following their lead, a number of other workers including, for example, F. N. Frenkiel, A. Skumanich, and M. S. Uberoi, erroneously concluded from actual photometric measurements that the brightness fluctuations were random and that a cellular pattern did not exist.

Since this time much progress has been made in understanding real turbulent fluids as opposed to the idealized models envisaged by Kolmogoroff and other early workers. Motions in the solar photosphere do not admit the simple description implied by the early development of statistical theories of turbulence. As we shall explain in Chapter 3, the existence of a well-defined cellular convective pattern does not exclude the presence of a smaller-scale turbulence which is produced by the shearing motions within the fluid.

Under the circumstances then prevailing, it is hardly surprising that by the mid-1950s there was widespread confusion among solar physicists as to the true appearance of the photosphere. Some workers remained convinced of the reality of the granulation and, as we shall see in the next section, made plans for improved observations, whereas others hesitated to admit its existence. A striking illustration of the state of affairs at the time is provided by the words of three very experienced observers at the McMath–Hulbert Observatory, R. R. McMath, O. C. Mohler, and A. K. Pierce, who began a paper published in *The Astrophysical Journal* in 1955 under the title 'Doppler shifts in solar granules' with the following sentence: 'In titling this note we do not necessarily imply the existence of granules as such, but, because of the ease of discussion and widespread use in the literature, we retain the term.' Fortunately, however, as we shall see in the next section, such confusion was destined to be short-lived.

1.6 Beginning of the modern era of high-resolution granulation observations

The beginning of the modern era of high-resolution granulation photography antedates the first successful spectroscopic observations of the granules by nearly a decade. It goes back to the early 1940s when the French astronomer Bernard Lyot (1897-1952), working at the Pic-du-Midi Observatory, pioneered the application of the cinematographic technique to the photography of the solar photosphere. By taking a large number of photographs at intervals that were short compared with the lifetime of the granules, he was able to obtain a few that were comparatively unaffected by seeing. Lyot himself went no further than to demonstrate the value of the technique, but some of his observations were used by C. Macris in 1953 to determine the lifetime of the granulation. The credit for introducing the cinematographic technique as a research tool in solar physics properly belongs to R. R. McMath (1891-1962) and his associates at the McMath-Hulbert Observatory, where motion-picture films of prominences and other chromospheric phenomena were obtained as early as 1933. Subsequent experiments with 'white-light' cinematography indicated that the new technique might also be ideal for recording changes in the photospheric granulation, but it was never systematically used for this purpose at the McMath-Hulbert Observatory. Following Lyot, cinematographic observations of the granulation were carried on by J. Rösch at the Pic-du-Midi with notable success; much of Rösch's work is described in the chapter that follows.

In 1957 the full power of the cinematographic method for granulation photography was convincingly demonstrated by R. B. Leighton at the Mt Wilson Observatory. Working in the early morning when the seeing conditions at the 18-m solar tower are at their best, he obtained sequences of high-quality photographs of the granulation using the full 30-cm aperture of the telescope. Like Strebel's photographs, Leighton's observations showed the granules as bright features of various shapes separated by narrow lanes of darker material, whose apparent width was comparable with the resolving limit of the telescope, 0″.4. The basic appearance of individual granules remained unaltered from photograph to photograph. As Leighton pointed out, these observations demonstrated conclusively that the granules were to be identified with *convection cells* and not with the eddies of any large-scale turbulence. Originally a physicist at the California Institute of Technology, and later a member of the staff of the Mt Wilson and Palomar Observatories, Leighton went on to develop ingenious new techniques for 'photographing' magnetic and velocity fields on the surface of the Sun, which led to the discovery of the oscillatory velocity field (Section 2.5.4) and the supergranulation (Section 2.7).

The next development took place in Australia when, in 1957, the Physics Division of the Commonwealth Scientific and Industrial Research Organization brought into operation a 13-cm photoheliograph specifically designed for high-resolution cinematography of the solar photosphere and sunspots. The observations obtained with this instrument were of sufficient quality to enable the development of individual granules to be followed from photograph to photograph. It was found that the majority of granules show comparatively little systematic change over most of their lifetimes. This stability strongly supported Leighton's conclusion that the motions within the granules themselves are basically laminar rather than turbulent, and, therefore, the view that the granules were to be identified with convection cells rather than 'turbulent eddies'. Moreover, the Australian observations, while not capable of showing granules as small as the very bright elements predicted by Richardson and Schwarzschild (Section 1.5), revealed no strong correlation between the brightness and size of individual granules: in fact, it was found that bright granules are just as frequently larger as smaller than average (see Section 2.3.5).

Parallel to the development of improved methods of observing the granulation from the ground, a radically different approach to the problem was being pioneered in England and France by D. E. Blackwell, D. W. Dewhirst, and A. Dollfus. Following earlier unsuccessful experiments from an aircraft flying at 6.7 km, these workers constructed a solar telescope for operation from a manned balloon and carried out several flights to heights of 6.1 km. Their balloon photographs were much better than those obtained at random from the ground but were nevertheless inferior to the best ground-level photographs. However, the manned flights did serve the useful purpose of showing that a balloon telescope needed to be operated at greater heights if the residual effects of atmospheric seeing were to be entirely avoided.

Meanwhile, in the USA a team of individuals and organizations under the direction of Martin Schwarzschild was building a 30-cm balloon-borne solar telescope for automatic operation in the stratosphere, a complex and expensive undertaking which became known as 'Project Stratoscope I'. The original specific aim was to test Richardson and Schwarzschild's prediction of the existence of small, very bright granules - hitherto unresolved - with which they claimed the most energetic 'turbulent elements' of the photosphere were to be identified. In the latter part of 1957 several successful flights were carried out at heights in the vicinity of 24 km, where 96% of the Earth's atmosphere lay below the telescope. During these and later flights granulation photographs of unsurpassed definition were obtained, one of which is shown in Fig. 2.1. However, as the reader may see for himself, they show no predominance of small, very bright granules and thus, by an ironic twist of fate, directly contradict Richardson and Schwarzschild's original prediction! Instead, the appearance of the solar surface is that of an irregular pattern of polygonal bright granules, mostly 1-2″ in diameter, separated by narrow lanes of dark material. This is of course identical to the picture provided by the ground-based observations of Strebel and Leighton, but to Schwarzschild, as he generously admitted, the pattern came as a complete surprise.

In view of the developments described above, the year 1957 may be taken as marking the beginning of a new era in the study of the solar granulation. Since then tremendous advances have been made in our observational knowledge of its properties, while theoreticians have devoted much attention to the problem of elucidating the nature of the convective processes responsible for its origin. The results of all this work are described in the succeeding chapters of the book.

1.7 Chronological summary

1801 W. Herschel uses the term 'corrugations' to describe the mottled appearance of the solar disk.

1862 Announcement of Nasmyth's 'willow-leaf' pattern.

1864 Dawes introduces the term 'granule'.

1866 Publication of a paper by Huggins ends controversy over Nasmyth's 'willow-leaves'.

1877 Granulation successfully photographed by Janssen.

1896 Publication of Janssen's collected observations.

1908 Hansky estimates mean lifetime of granules to be about 5 min.

1914 Publication of Chevalier's collected observations of the granulation.

1930 Unsöld attributes the origin of the granulation to convection currents in the hydrogen ionization zone.

1933 Announcement of Strebel's discovery of the polygonal shapes of the granules.

1933 Siedentopf formulates a theory of the granulation based on Prandtl's mixing-length theory of turbulent convection.

1936 H. H. Plaskett identifies the granules as Bénard-type convection cells.

1949 Richardson obtains spectra showing the Doppler shifts of the granules.

1950 Richardson and Schwarzschild identify the granules as the eddies of a large-scale aerodynamic turbulence.

1953 First reliable determination of granule lifetimes made by Macris.

1955 Various workers claim that the solar surface shows random brightness fluctuations, not a cellular pattern.

1957 Rösch publishes granule observations made at the Pic-du-Midi.

1957 High-resolution photoheliograph brought into operation near Sydney by the CSIRO.

1957 Leighton re-asserts convective origin of the granulation.

1957 Granulation photographed from a manned balloon.

1957 Project Stratoscope I yields granulation photographs of unsurpassed definition.

2

THE MORPHOLOGY, EVOLUTION AND DYNAMICS OF THE GRANULATION AND SUPERGRANULATION

2.1 Introduction

The appearance of the granulation in the central region of the solar disk is well illustrated in Fig. 2.1, which is an enlargement of a high-quality photograph obtained by M. Schwarzschild and his collaborators with a 30-cm balloon-borne telescope on 17 August 1959. The photograph shows that the granulation consists of a *cellular* pattern of bright elements on a darker background. The majority of the granules appear to have diameters in the range 1–2″ of arc (725–1450 km) and are separated by narrow dark lanes, whose apparent width often does not exceed a few tenths of a second of arc. In places, however, there are relatively large areas of dark intergranular material, which seem to result from the absence - presumably temporary - of one or more granules. These occasional dark regions were called porules by Rösch (1959), and are characteristic features of the granulation pattern; they should not be confused with *pores* (small sunspots with no penumbra), which are not only darker but also very much longer-lived (Bray & Loughhead, 1964, p. 69).

It is instructive to compare the appearance of the granulation on Fig. 2.1 with that on Fig. 2.3, which shows two photographs taken with ground-based telescopes. The first of these (Fig. 2.3(*a*)) was obtained by Bray and Loughhead on 11 June 1976 with a 30-cm refractor at the CSIRO Solar Observatory, Culgoora. It is of a slightly lower resolution than the stratospheric photograph, but the polygonal outlines of the granules and the narrowness of the dark lanes separating them are again well shown. The second photograph (Fig. 2.3(*b*)) was obtained by Bray and Loughhead on 10 January 1960 with the earlier Sydney 13-cm photoheliograph. On this photograph the granules have lost their polygonal appearance as a result of the lower resolution, although their diversity in shape and size is still apparent. Moreover, the dark lanes between the granules are now much wider and more diffuse. In this regard, it is important to emphasize that in the velocity measurements subsequently described in this

Fig. 2.1. Photograph of the photospheric granulation obtained by M. Schwarzschild and his collaborators with a 30-cm balloon-borne telescope on 17 August 1959. (By courtesy of Project Stratoscope of Princeton University, sponsored by ONR, NSF and NASA.)

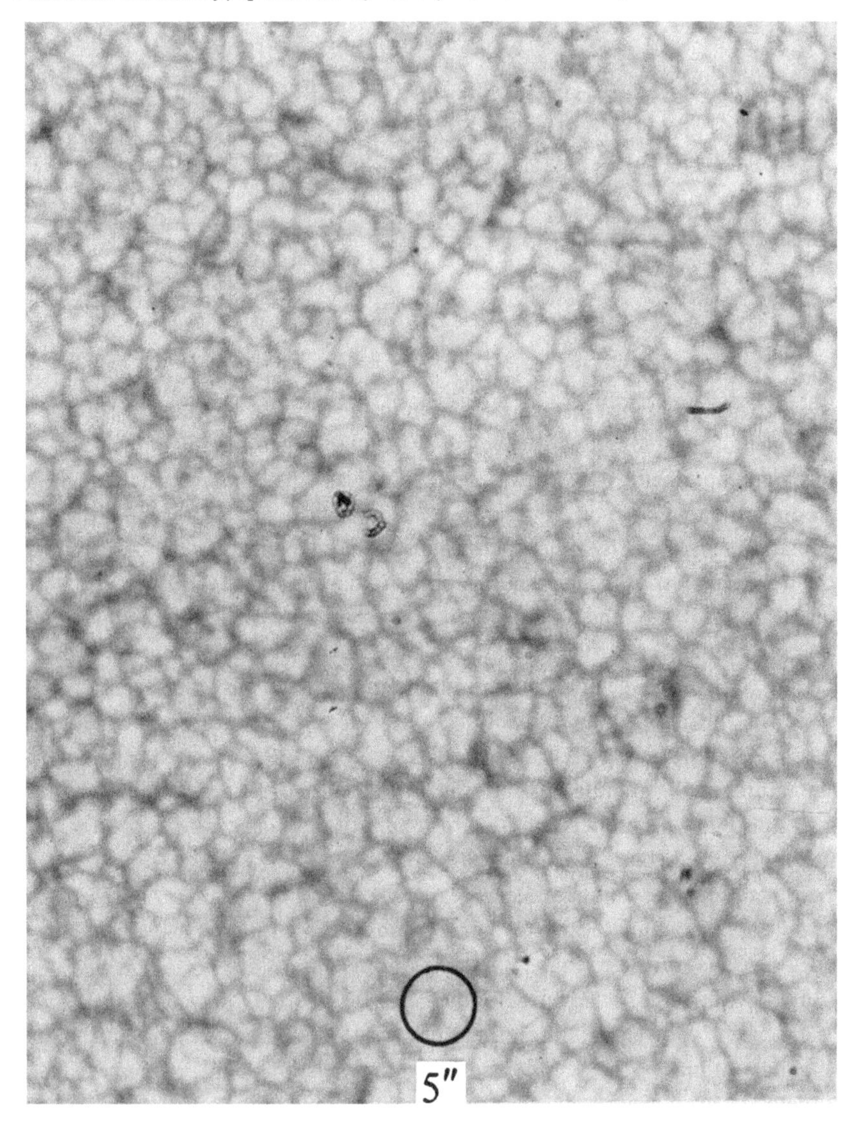

Fig. 2.2. Photograph of the photospheric granulation obtained by V. A. Krat and his collaborators with a 50-cm balloon-borne telescope on 30 July 1970. (By courtesy of V. N. Karpinsky, Pulkovo Observatory.)

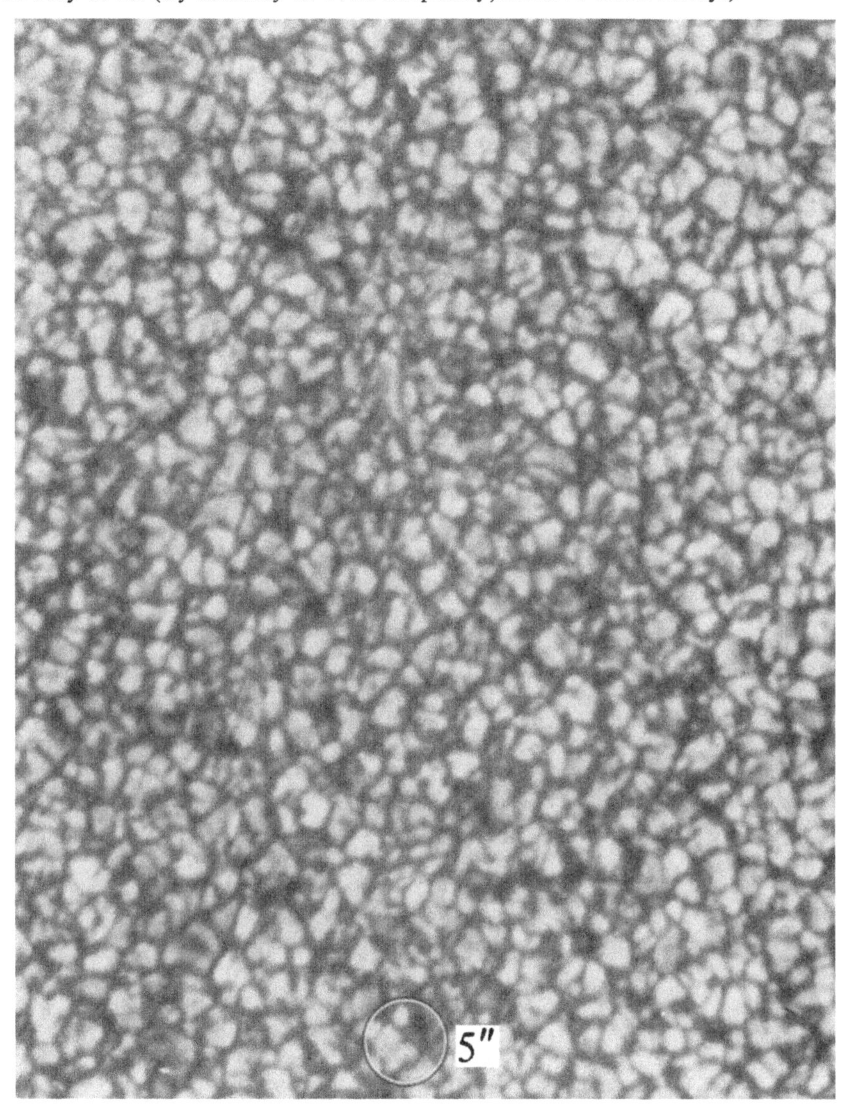

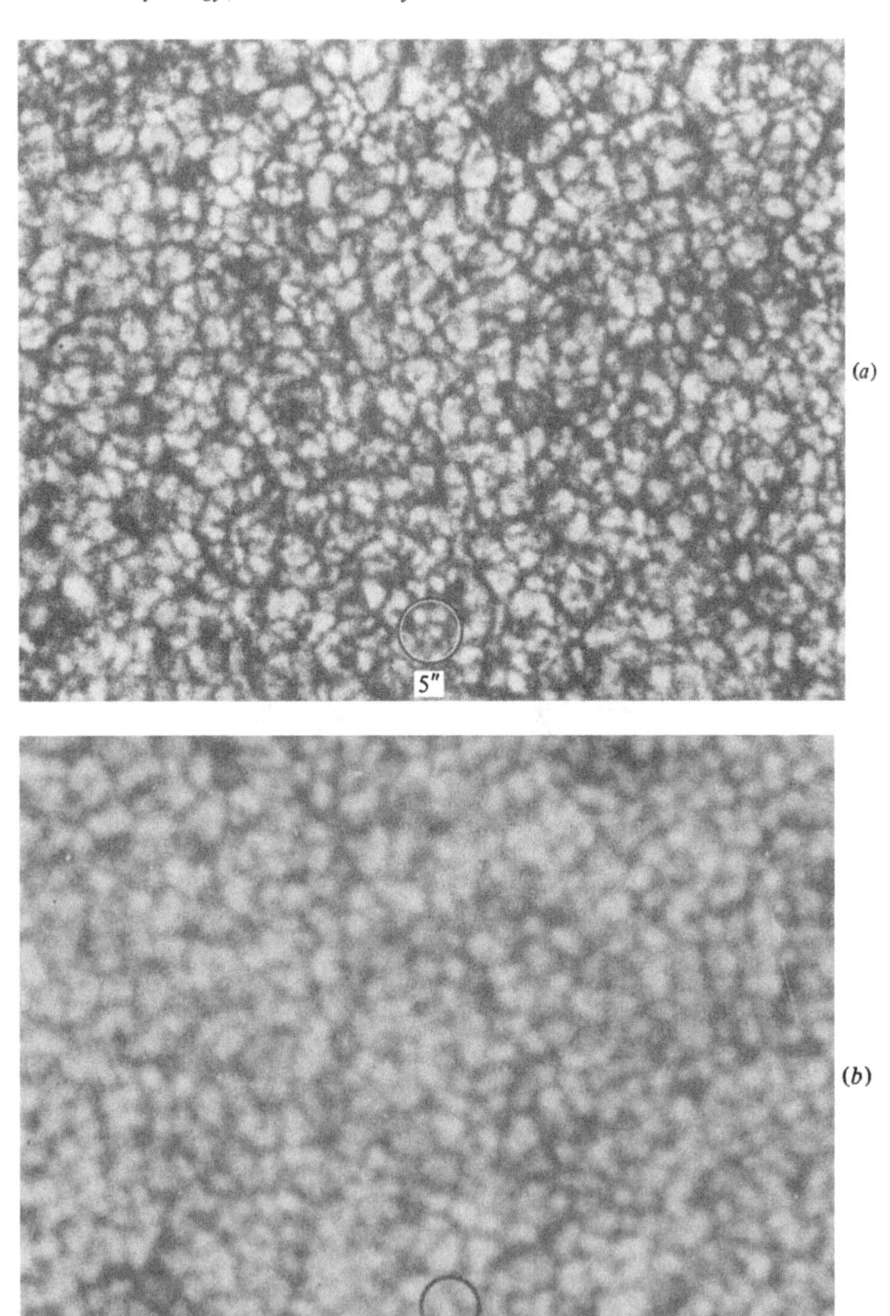
5″
(a)
5″
(b)

chapter, the effective spatial resolution is mostly comparable to that of Fig. 2.3(*b*) rather than to that of Fig. 2.1 or Fig. 2.3(*a*).

Our aim in the present chapter is to give a critical and comprehensive account of current observational knowledge of the properties of the granulation pattern as a whole and of the individual granules. We shall be concerned solely with the photospheric granulation and the supergranulation; we refer the reader to the works cited in Table 2.13 for discussions of other fine structures of the solar photosphere. Moreover, little or no attempt is made to give a theoretical explanation of the various observed features, the whole question of the origin of the granulation and the interpretation of its properties being dealt with in Chapter 5.

As a prelude to our discussion of the observed properties of the granulation, we describe some of the difficulties attendant on the attempts to measure these properties. The Earth's atmosphere sets a definite limit to our ability to see objects of the apparent size of the photospheric granules and to the accuracy with which we can determine their properties. We shall be confronted with these limitations in the subsequent discussions, so their nature and origin (described in Sections 2.2.1 and 2.2.2), as well as the means of overcoming them (Section 2.2.3), should be borne in mind.

We begin our discussion of the photospheric granulation by describing the properties of the individual photospheric granules (Sections 2.3.1 to 2.3.7) dealing successively, among other things, with their shape, diameter, cell size, contrast, lifetime, and evolution. In Section 2.4.1 we consider the important question of how close to the limb the granulation can still be perceived, and hence how far the convection currents responsible extend into higher layers of the photosphere. The observations show unmistakably that the granules are seen well into the upper region of the photosphere which, according to the theory of the hydrogen convection zone (see Sections 4.2.2 and 4.2.4), is convectively *stable*; the evidence for velocities associated with granules in the convectively stable regions is discussed in Section 2.4.2. This phenomenon is akin to the overshoot ordinarily observed in naturally occurring convection (Section 3.4.5).

Fig. 2.3. The photospheric granulation.
(*a*) Photograph obtained by Bray & Loughhead with the 30-cm refractor of the CSIRO Solar Observatory, Culgoora, on 11 June 1976.
(*b*) For comparison, a photograph of lower resolution obtained with the Sydney 13-cm photoheliograph on 10 January 1960.
Both photographs are on the same scale; the diameter of the circle is 5″ of arc. Slight variations in seeing across (*a*) are typical of ground-based observations.

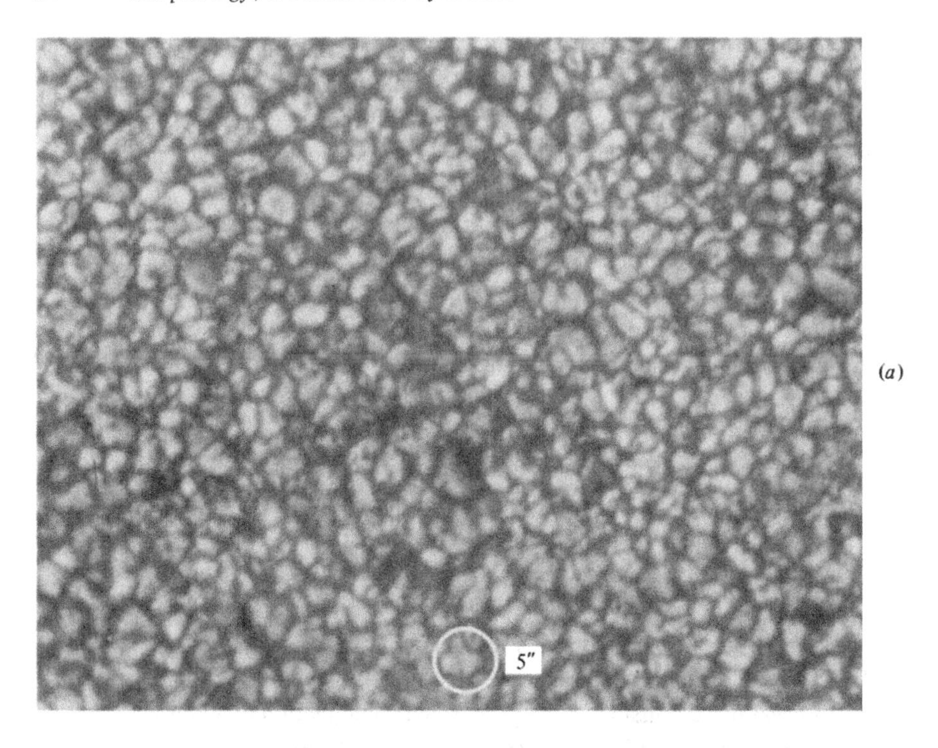

(a)

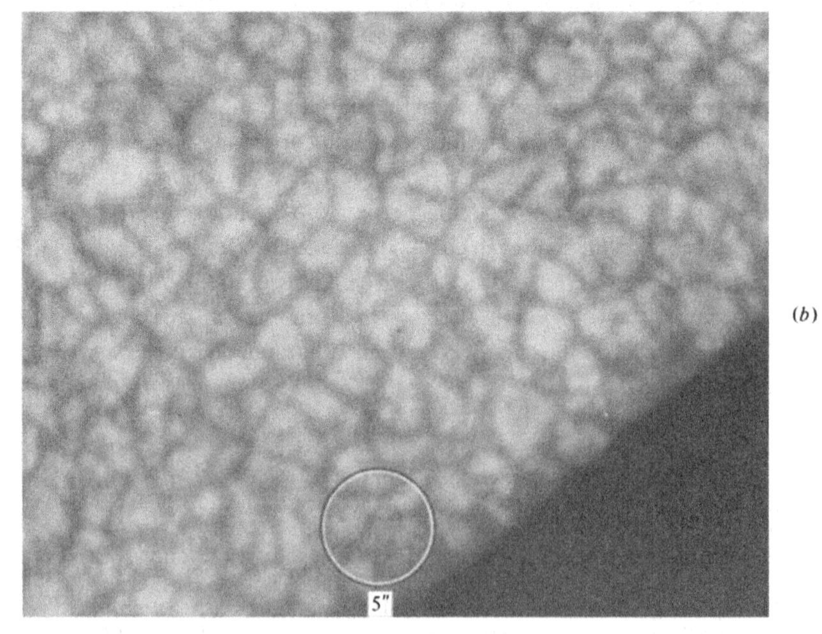

(b)

In Section 2.5 we turn to the properties of the granulation pattern as distinct from those of the individual granules. The stochastic nature of the appearance and evolution of individual granules encourages a statistical analysis of the granulation field. The brightness distribution is dealt with in Sections 2.5.2 and 2.5.3, where we look at the root mean square (rms) values and mean spatial scale, together with their variations with height, heliocentric angle, and wavelength.

The velocity distributions are treated in Section 2.5.4. The interpretation of the velocities measured in the photosphere is complicated by the presence of a large-scale oscillatory field, the velocity at any given point varying in an almost sinusoidal fashion with a period close to 5 min. It is believed that this oscillation is a global acoustic mode of the Sun; it has no connexion with convective processes. The distinction is clearly seen by consideration of the correlations between intensity and velocity described in Section 2.5.5. Only when the oscillations are filtered out does a high correlation between bright features and upward velocities and between dark features and downward velocities appear - as is to be expected for a convective process.

In Section 2.6 we consider the question of whether there is any relationship between the granulation and the magntic fields known to be present in the photosphere. Despite great difficulties, observations designed to resolve the question appear to indicate that any correlation between the dispersed magnetic fields outside activity regions and the granulation is either weak or entirely absent. Even within activity centres, where the mean field strengths are higher, the granulation generally appears to remain unaffected. Definite disturbances seem to occur only in restricted regions where it is thought that strong magnetic fields are concentrated.

Finally, in Section 2.7 we look at the large-scale and comparatively well-ordered system of horizontal motions to which R. B. Leighton drew attention in 1960. Their appearance is highly suggestive of the flow patterns observed in cellular convection. Although the velocities are mainly horizontal and the cell size is much larger than that of the photospheric granulation, the velocity field

Fig. 2.4. Photographs of the photospheric granulation obtained with the 40-cm vacuum Newtonian reflector at Izaña on
(*a*) 31 July 1979. Slight variations in seeing are apparent across the frame but the best resolution is equal to that of stratospheric pictures (by courtesy of H. Wöhl);
(*b*) 30 June 1973 during a partial solar eclipse. The intensity profile of the lunar limb allows the blurring of the image by instrument and seeing to be measured directly. The irregular polygonal shapes of the granules and the linearity of the intervening lanes are well brought out with the higher magnification of this photograph. (By courtesy of W. Mattig.)

nevertheless bears a certain resemblance to the ordinary granulation and was accordingly named by Leighton the 'supergranulation'.

For the convenience of the reader, the chapter ends with a table summarizing the most important quantitative data concerning the main physical properties of the photospheric granules. For comparison, the table also includes the corresponding data for other structures seen in the photosphere.

A table of the principal symbols employed in this chapter will be found in Chapter 3, Table 3.1.

2.2 High-resolution observing methods

It will be clear from Chapter 1 that observations of the granulation lie at the limit of those technically feasible with both ground-based and balloon-borne telescopes. The angular size of a typical granule is of the order of one to two seconds of arc. To see the internal structure and the intergranular lanes requires a subarcsecond resolution, and such spatial resolution is very hard to achieve. Balloon-borne telescopes have tended to be too small in aperture, whilst ground-based telescopes suffer from image degradation, known as solar seeing, produced by the Earth's atmosphere. We shall discover that we still rely very heavily on observations from the ground, so the question of solar seeing is fundamental to the problem of obtaining high-resolution observations.

Accordingly, we begin with a brief description of solar seeing and its causes in Section 2.2.1, while the connexion between the Earth's atmosphere and the quality of solar images is reviewed in Section 2.2.2. The various strategies for choosing sites that provide optimum conditions for high-resolution observations, which have grown out of our developing understanding of the causes of seeing, are also described in Section 2.2.2. In Section 2.2.3 we look at two means of coping with the seeing problem. To a large extent, it can be avoided by observing high in the atmosphere from balloons or above it from satellites. This is an expensive, uncertain, and short-lived solution to the problem; therefore much effort is being put into the development of techniques of improving ground-based observations by correcting for the inevitable atmospheric distortion.

2.2.1 Solar seeing: blurring and image motion

Light emanating from a point of the solar surface spreads as a spherical wave. By the time it reaches the Earth the curvature of this wave is so small that it can be neglected, and the wave front can be treated as plane. Then, however, the light wave enters the Earth's atmosphere in which temperature perturbations associated with dynamical disturbances – the result of the usual processes of convection and turbulence – cause the index of refraction to fluctuate. These random disturbances progressively distort the originally plane wave front as it passes through the atmosphere. The wave front finally arrives at the telescope possessing a constantly changing corrugation due mainly to small-amplitude phase retardations (see Stock & Keller, 1960; Meinel, 1960). The intensity of the received wave is scarcely modulated if gross effects, of clouds for example, are excluded.

If the aperture of the telescope is large compared to the size of the atmospheric disturbances, the incident wavefront can be pictured as the product of a plane wave and a noise function. The wave front in the focal plane is then determined by the convolution of the Fourier transform of the plane wave (i.e. the Airy function) and that of the noise function. The image therefore consists of a number of diffraction images of the aperture whose positions are determined by the peaks of the Fourier transform of the noise. These diffraction images combine to form an image of the original point on the solar surface whose overall diameter is typically of the order of a second of arc.

Since the wave front distortion is constantly changing, the number and location of these diffraction images change also; but the pattern will be effectively frozen if the exposure time is short compared with the time scale of the atmospherically induced fluctuations. Under these circumstances, the individual diffraction images, known as *speckles*, will not be degraded. The reconstruction of diffraction-limited images of point sources using these speckles is a rapidly developing technique in astronomy (see Weigelt, 1978).

Difficulties arise in application to solar observations because the Sun is, of course, not a point object. Speckle techniques can be applied only when the field of view is smaller than the average separation of the speckles. Since the total spread of the pattern is only a second of arc or so across, this is an unacceptable restriction. A field of view large enough to encompass even a single granule inevitably produces overlap and smearing of the speckles, which must be taken into account in some manner.

If the number of speckles is small - as is usually the case for a telescope with a small aperture admitting only a limited number of elementary waves - there will be little smearing and the image will appear sharp. But the statistical fluctuation in the positions of the speckles will be large, and the centre of gravity of the composite image will move around from instant to instant. This movement is known, naturally enough, as image motion. If the object is extended, light from different parts of it will follow different paths through the atmosphere. This leads to differential motions within the image, or image distortion (Rösch, 1963). A telescope of large aperture, on the other hand, admits more elementary waves and produces a composite image with correspondingly less motion but more blurring. In general, effects of both image motion and blurring are present in any telescopic image and together form the 'seeing'.

The average degree of blurring and image motion in any given instrument tends to be correlated although the relationship breaks down when instantaneous values are compared. The general correspondence is illustrated in Fig. 2.5 which compares the rms image motion with the median value of the full width at half maximum (FWHM) of the blurring function, measured at the McMath telescope of the Kitt Peak National Observatory (Brandt, 1976).

The exposure times required for photographic spectrophotometry are usually much longer than the time scale of the atmospheric turbulence, and a temporal averaging will therefore occur. Each point on the Sun will produce a set of speckles whose combined blurring and integrated motion produces a two-dimensional image profile whose envelope is known as the point-spread function (PSF); the so-called 'seeing disk' is a rough measure of the width of this function. The recorded image is then the convolution of the ideal, undegraded, solar image with this point-spread function. The half-width of the seeing disk can range from several seconds of arc down to about 0".2 under the most favourable circumstances.

Fig. 2.5. Correlation between image motion and image blurring. Ordinate: rms image motion; abscissa: median value of the full width at half maximum (FWHM) of the blurring function measured over 60 s intervals. Different symbols indicate observations on four days at the 92-cm McMath telescope, Kitt Peak National Observatory (after Brandt, 1976).

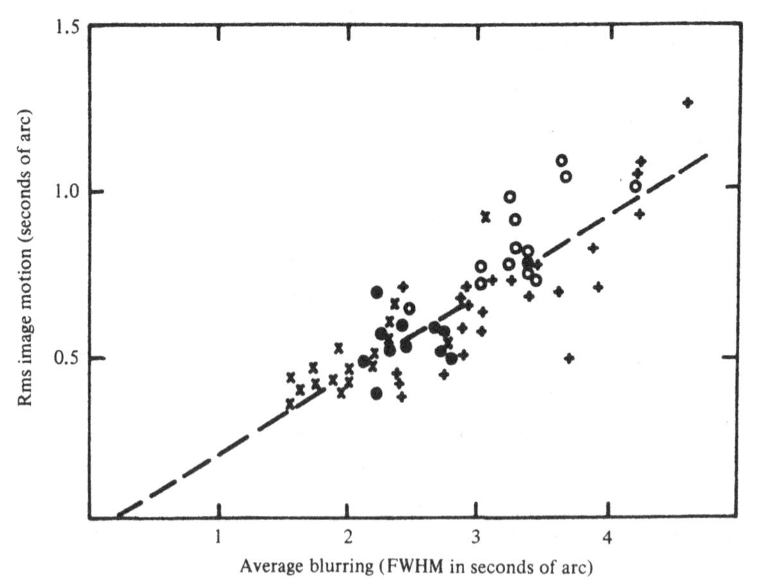

The measurement of the point-spread function, which would allow the reconstruction of the ideal image from the recorded image, is a very difficult task since it varies with the seeing not only from moment to moment but also across the field of view (see Section 2.2.3). Not surprisingly, observers have sought to minimize the effects of solar seeing by choosing sites with the least atmospheric disturbance.

2.2.2 Solar seeing and the Earth's atmosphere

The dynamical disturbances of the atmosphere responsible for solar seeing may have either a convective or a turbulent origin. Convection arises mainly from the heating of the ground surface by the Sun and the convection currents can rise several thousand metres above the ground. Turbulence is produced by unstable shear layers; these can occur throughout the troposphere, but most markedly in the tropopause at heights of 8000–16 000 m depending on latitude and season. Due to the decreasing air density, the upper atmospheric layers are generally insignificant - except possibly when a jet stream is present - as sources of degradation compared with the layers immediately surrounding the telescope (see Barletti *et al.*, 1977).

Our picture of the convective layers of the Earth's lower atmosphere owes much to the pioneering work of the Australian meteorologists under C. H. B. Priestley (Priestley, 1959). Over flat terrain they distinguished three distinct regimes: a *forced* convection regime extending a metre or so above the ground in which heat is carried upwards by turbulent motions generated by surface wind shear; a *composite* convection regime extending up to some tens of metres where the motions carrying the heat are set up partly by the wind and partly by buoyancy forces; and a *free* convection regime where buoyancy forces are dominant.

Fig. 2.6 illustrates the nature of the temperature fluctuations associated with the three separate convective regions (Webb, 1964). In the forced convection region the temperature traces have the random appearance of purely turbulent fluctuations. In the composite region, on the other hand, in addition to the random fluctuations there are discrete periods of marked temperature enhancement ('bursts'), which become more prominent with increasing height. Throughout much of this region and in the free convection zone above it, many of the bursts are recorded almost simultaneously by instruments placed at different heights showing that they have the form of *buoyant plumes* extending upwards to a considerable height. However, the temperature bursts are generally registered slightly earlier at higher levels, indicating that the plumes are tilted as they are carried along by the wind. In addition, there is generally a decrease in the duration of the recorded temperature bursts in the upper part of the free

Fig. 2.6. Convection and seeing over flat terrain.
(*a*) Typical measured temperature fluctuations within the three separate convection regions in the boundary layer of the Earth's atmosphere resulting from heating of the ground by the Sun (adapted from Webb, 1964).
(*b*) Four typical examples of moments of good seeing, demonstrating the striking suddenness of their occurrence. The dashed line indicates the 0.8 volt level at which the seeing becomes promising for high-resolution observations (see text). Most good moments are shorter than those shown, lasting no longer than 2–3 s (Loughhead & Bray, 1966).

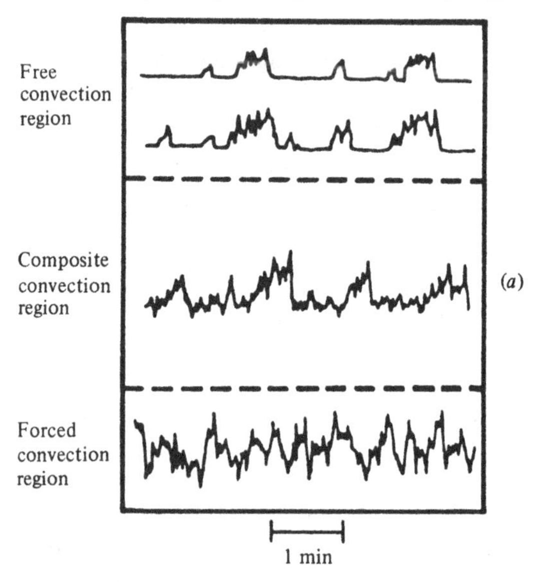

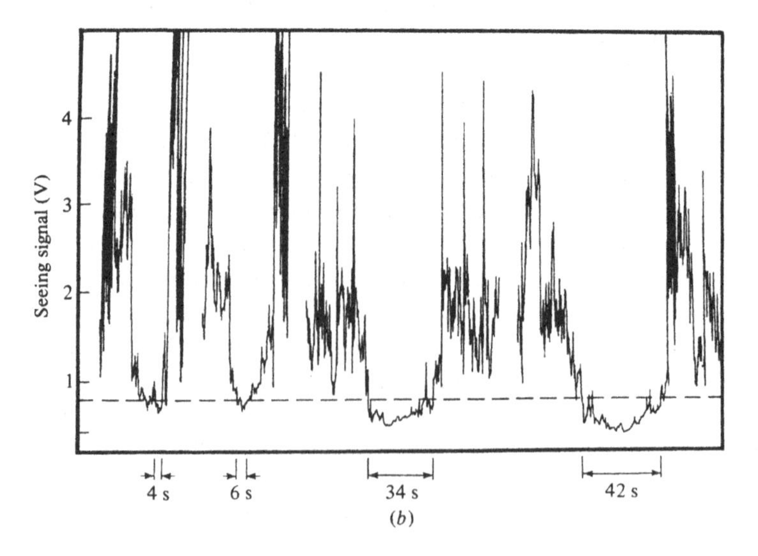

convection region, implying that the width of the plumes tends to *decrease* with height.

In the free convection region, as illustrated in Fig. 2.6(*a*), the temperature traces show a well-marked alternation of disturbed and quiet periods. The disturbed periods mark the passage of the rising convective plumes, in which the temperature fluctuations are of the order of 1-2 K (Webb, 1964). On the other hand, the quiet periods are associated with much more uniform descending air outside the plumes, the magnitude of any residual temperature fluctuations still present appearing to be less than 0.1 K (Priestley, 1959). The quiet periods may last from a few seconds up to a minute or more, and thus bear a marked resemblance to the sudden moments of good seeing recorded by the image-motion monitor developed by Bray, Loughhead & Norton (1959).

Fig. 2.6(*b*) shows the records obtained on four fairly typical occasions when, briefly, the seeing signal dropped below the level, shown by the dashed line, at which the situation becomes 'promising' from the point of view of 1-2″ of arc direct solar photography (0.8 V). (There is a good correlation between the values of the seeing signal, expressed in terms of an arbitrary voltage scale, and the resolution shown by direct photographs of the photosphere taken at the same time.) All four examples provide striking demonstrations of the characteristic *intermittency effect* in solar seeing. When a quiet period occurs the seeing signal declines with great rapidity, and afterwards increases to a more average value equally quickly. It was the resemblance between the temperature fluctuation records and the seeing monitor traces which led Bray and Loughhead – following a suggestion by Priestley – to conclude that moments of good seeing occur when the line of sight momentarily passes through the quiescent regions surrounding the rising convective plumes (Bray & Loughhead, 1961; Webb, 1964). These considerations governed the design of a larger (30-cm) solar refractor erected by the CSIRO in Australia in north-west New South Wales, at Culgoora, some 600 km from Sydney (Loughhead, Bray, Tappere & Winter, 1968). The major factor influencing the seeing conditions is the wind speed. With increasing wind speed the height to which the deleterious forced convection makes itself felt increases (Webb, 1964). A similar thickening of the forced convection layer occurs if the terrain around the telescope is at all rough (a flat snow layer is optimal!).

If longer periods of good seeing than those provided by the chance alignment of downdraft and telescope are required, a radically different type of site must be sought; then the picture of the convective layers is by no means so simple as that described above. The need for a plausible picture of the more general situation was recognized during the search for a new solar observatory site in Europe undertaken by the Joint Organization for Solar Observations (JOSO).

The assessment of a site is mainly handicapped by the lack of systematic objective information concerning the general relationship between seeing and the conditions responsible for it. As described above, quantitative measurements of seeing quality were initiated in 1959 by the Australian observers. From these measurements and the confirmatory work of Kallistratova (1970) in the Russian steppes, stems our understanding of the seeing conditions at sites located on extended plains. Since then, instruments for monitoring solar seeing have increased in sophistication and scope and can now record both image motion and blurring (Brandt, 1970). But there have been very few extensive studies of the conditions at observatories with a more complicated topology. Observations of the seasonal variation of seeing are apparently available only for the earlier CSIRO site (Loughhead & Bray, 1966) and, to a more limited extent, for the Canary Island sites (Brandt & Wöhl, 1982). In the absence of more data, an empirical model of the general connexion between site and seeing is not even in prospect and we must be content with broad generalizations (Brandt, 1976; JOSO Annual Reports 1970–9).

One accepted principle of site selection invokes the stabilizing effects of large masses of water. These heat up more slowly than the ground and exchange less heat with the atmosphere. Hence the air above the water tends to be more stably stratified. In order to exploit this, an instrument may be sited either within a large extent of water as at the Big Bear Observatory (a facility of the Hale Observatories) situated in the hills some 80 km east of Los Angeles, or on

Fig. 2.7. Mean diurnal variation of the best seeing, measured by the smallest FWHM of the blurring function, at Capri (dashed) and Izaña (full). The opposite trends typical of coastal and high-altitude sites are clearly evident (Brandt, 1976).

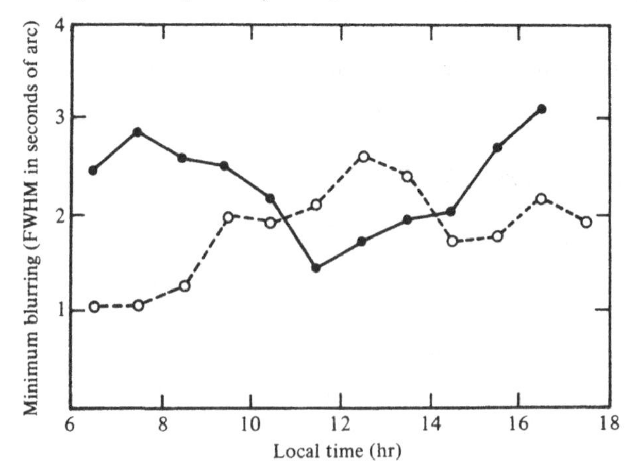

the coast where the higher temperatures over land draw a stable air flow in from the sea. At such sites, as at sites located on plains, the best seeing occurs at midday, when the line of sight through the atmosphere is shortest. This behaviour is demonstrated in Fig. 2.7 for the Capri station of the former Fraunhofer Institut - now Kiepenheuer Institut - (see Brandt, 1976). On Capri, though, orographic disturbances caused by the hilly nature of the island are very marked. These sites seem capable of providing quite prolonged periods of reasonably good, but rarely excellent, seeing conditions.

In order to improve on this, it would seem better to avoid as much as possible of the disturbing atmosphere by going to high altitudes. At Sacramento Peak Observatory, at an altitude of 2800 m in the Sacramento Mountains of New Mexico, USA, the best, and sometimes excellent, seeing tends to occur quite early in the morning, long enough after sunrise for the path through the atmosphere to have shortened but before the atmospheric convection has had time to develop. The diurnal variation of seeing quality at high-altitude sites is almost a mirror reflection of that at coastal sites, as may be seen in Fig. 2.7 from the examples of Capri and Izaña. Fig. 2.8 shows a more detailed comparison of the diurnal variation of seeing at two Canary Island sites, Izaña on Tenerife and Roque de los Muchachos on La Palma (both at altitudes of 2400 m), for the month of July, 1979 (Brandt & Wöhl, 1982). The curves show the fraction of time for which the blurring was less than $1\overset{''}{.}2$ (FWHM), as measured with 40- and 45-cm telescopes. Again we find the best conditions early in the morning, before strong local convection currents form. Also shown in Fig. 2.8 is a schematic curve depicting the magnitude of the air temperature fluctuations measured at Izaña multiplied by a function describing roughly the effect of the diurnal change in path length through the atmosphere (see Young, 1974). The close agreement between the two curves indicates that this product provides a plausible criterion for assessing the image quality at such sites.

An additional advantage offered by some mountain sites is their location above the atmospheric inversion layer. In maritime regions off the west coasts of continents the atmospheric temperature becomes subadiabatic at about 1500 m (dependent on season) above sea-level. Convection arising from the ground below the inversion level is unable to penetrate into the stably stratified region above. A mountain-top site located above this level is shielded from the convection currents arising below. Under favourable wind conditions, the convection currents originating in the immediate vicinity can be suppressed or blown away from the telescope's line of sight, which leads to extraordinarily good conditions for lengthy periods of time. Such conditions sometimes occur at the Pic-du-Midi Observatory in the Pyrenees which, though not itself a maritime region, comes occasionally under the influence of the Atlantic inversion.

The site chosen for the new combined German–Spanish Observatory at Izaña will, it is hoped, benefit more often from this inversion. White-light films taken there during the site testing campaign in 1979 with a 40-cm evacuated reflector show many frames of almost stratospheric quality. Fig. 2.4(*a*) is an example.

Once an optimal observing site has been identified, it is necessary to ensure that the disturbances caused by the instruments themselves do not negate the effort expended in site selection. Even if the telescope is raised above the layer of forced convection, the thermal disturbances within telescope, spectrograph, and dome can be just as important as the orographically induced effects. The solutions to these problems are very varied indeed, ranging from evacuated telescopes in double towers (to protect them from wind shake) to open telescopes standing in a free air stream. Some different systems were described in the first edition of this book, but regrettably there is still no known optimal solution.

Fig. 2.8. Diurnal variation of the best seeing at two high-altitude sites, Izaña and Muchachos in the Canary Islands (July, 1979). The full (Izaña) and long-dashed (Muchachos) lines give the fraction of the sample time during which the FWHM of the blurring function remained less than 1$''$.2 of arc. The short-dashed line is a schematic curve depicting the combined effects of the diurnal variation of the average atmospheric temperature fluctuations and the varying length of the light path (Brandt & Wöhl, 1982).

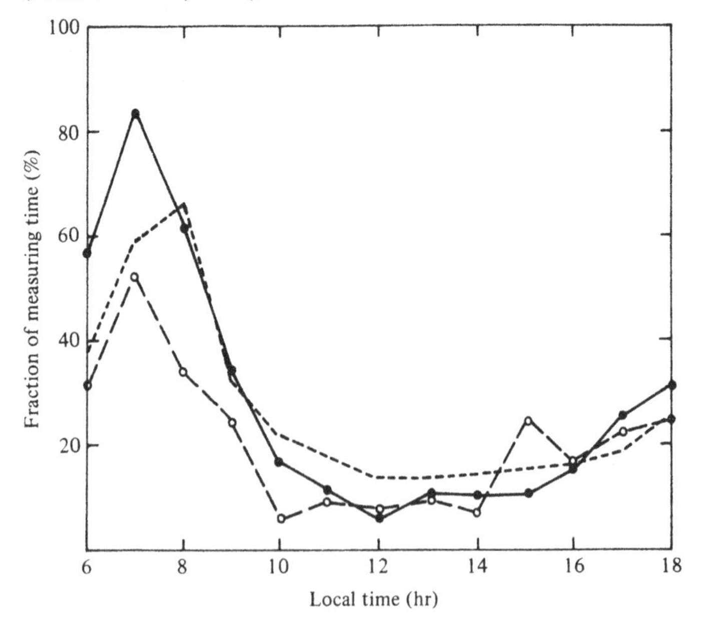

Brandt (1976) draws cautious conclusions from a limited survey of observatories, viz. Capri, Izaña, Kitt Peak and Sacramento Peak. His results are summarized in Fig. 2.9, which shows for each site how often the quality of the seeing, measured here by the minimum value of the blurring within a 60 s interval, falls below any given value. The superiority of the mountain-top sites equipped with evacuated telescopes (Izaña and Sacramento Peak) is apparent but the statistical basis for the comparison is still weak.

2.2.3 Elimination of solar seeing effects

Short moments of good seeing occur *comparatively* frequently because of the intermittency effect discussed above. These allow high-resolution white-light photographs to be obtained with some regularity, but the steadier conditions of good seeing necessary for spectrographic sequences are much more rarely fulfilled. The granulation lies at the very limit of the spatial resolution obtainable from the ground and great patience is required to obtain high-

Fig. 2.9. Frequency of moments of good seeing at four observatories. Within each 60 s interval of observation the smallest value of the FWHM of the blurring function was taken as a measure of the image quality. Plotted is the (cumulative) fraction of the number of intervals for which the image quality was better than any chosen value. The total length of the observations varied from 6 to 112 hr (Brandt, 1976).

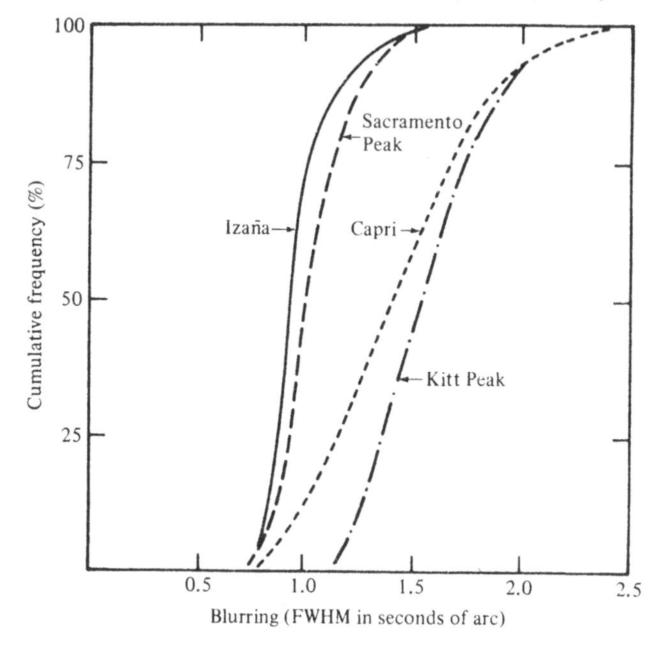

quality observations. This frustrating circumstance has encouraged attempts to improve the resolution by two methods. The first avoids the atmospheric distortion by utilizing remotely controlled instrumentation on platforms lifted above the atmosphere; the second accepts the distortion but seeks to compensate for it.

Heading the first category is the Stratoscope experiment already described in Chapter 1. It was flown in both 1957 and 1959 and delivered white-light photographs that revealed the morphology of the solar surface in unprecedented detail.

The success of Stratoscope I inspired two subsequent experiments in both of which a spectrograph was attached to the telescope so that spectrophotometry could be added to the morphological analysis. The Spektrostratoskop balloon-borne experiment of the Fraunhofer Institut was flown on 17 May 1975 (Mehltretter, Mattig & von Alvensleben, 1978). Observations were obtained over a period of 6 hr at a height of 28 km. The 32-cm telescope provided an image scale of 0.45 mm arcsec^{-1}; the spectrograph had a dispersion of 180 mm nm^{-1} and was equipped with a polarization analyser for the measurement of magnetic fields.

The Soviet Solar Stratospheric Observatory (SSSO) made two flights during which the granulation was observed. On 30 July 1970 it carried a 50-cm telescope equipped with a spectrograph with a dispersion ranging from 12.5 mm nm^{-1} in the red to 19.0 mm nm^{-1} in the blue (Krat, Karpinsky & Sobolev, 1972). One of the best white-light photographs obtained during this flight is shown in Fig. 2.2. On a later flight, on 20 July 1973, a 1-m telescope was employed (Sobolev, 1975). Neither the German nor the Soviet experiments have been without their photometric problems but each has yielded sequences of a quality and *homogeneity* unequalled from the ground. The results will be reviewed extensively in the individual sections below.

Vehicles other than balloons have not, as yet, been utilized for granulation observations. Rocket flights are quite unsuited as they are short-lived and can carry an instrument of only limited size. Restricted size and inadequate spatial resolution have also characterized satellite instrumentation to date. In the late 1980s, the flights of NASA's Space Shuttle may offer the first real opportunity for definitive undegraded observation of the granulation, provided plans go ahead for a large 'Solar Optical Telescope' (SOT) of 1.25-m aperture equipped with a powerful spectrograph for use in the visible region of the spectrum.

In the future, as now, we may expect many valuable observations to stem from ground-based observatories. The best photographs obtained at some of these observatories, such as those illustrated in Figs. 2.3(*a*), 2.4 and 2.25, are almost equal to those of the balloon-borne telescopes, illustrated in Figs. 2.1 and

2.2. This judgement may be verified by reference to the various measurements of the rms brightness fluctuation collected in Table 2.9. Also more effort is being made to avoid the need to wait on chance occasions of exceptional atmospheric quietness.

The most ambitious aim is to correct the distortion of the wave front in the course of the observation. If a point-like, high-contrast object can be found in the field, the angular fluctuations of the wave front stemming from it can, in principle, be sensed and used to compensate the main image by suitably distorting an optical element in the telescope. This can be achieved by flexing a thin mirror, a so-called rubber mirror, or by tilting the individual mirrors of a multi-element telescope relative to one another so that the point images are brought into coincidence (see collections of papers devoted to adaptive optics edited by Fried, 1977, and Dunn, 1981).

This image-correction procedure is unfortunately of limited utility as far as the observation of spatially extended objects is concerned. The correction requires all elements of the image, arriving at different angles, to have passed through the same atmospheric disturbances. If that is not the case, the speckle patterns, and thus the point-spread function, will be different at different points of the image. The corrections will go progressively out of register as we proceed away from the reference point. The extent over which the point-spread function is sensibly constant is known as the isoplanatic patch. Both experiment and theory show that the instantaneous isoplanatic patch is, at most, only a few seconds of arc across (Young, 1974; Lohmann & Weigelt, 1979). Nevertheless, in recent years several workers (see Dunn, 1981) have pursued the application of this technique to the observation of photospheric fine structure.

A considerable improvement in image quality can be achieved by much simpler means. If just the motion of some high-contrast object is sensed, the whole image can be displaced by a servo-controlled tilting mirror so that the object is maintained in a fixed position. The image motion in the surrounding isoplanatic patch will then be removed; this can lead to quite striking improvements in image quality (Tarbell & Smithson, 1981).

A less ambitious aim is to effect the seeing correction not in real time but subsequent to the observation. Progress is being made in this direction as a result of improvements in the mathematical modelling of the connexion between atmospheric turbulence and astronomical seeing. This subject owes much to the pioneer work of Chandrasekhar (1952), Tatarski (1961) and Coulman (1965, 1969). In recent years the subject has also been studied by a French group at Nice, whose work has been comprehensively reviewed by one of its members (Roddier, 1981). Theories based on the assumption of isotropic turbulence in the terrestrial atmosphere predict that the spatial distribution of the mean quadratic difference in the refractive index n between two points depends only on their separation; thus

$$\overline{(n(\mathbf{x}_1)-n(\mathbf{x}_2))^2} = C_n^2|\mathbf{x}_1-\mathbf{x}_2|^{2/3}, \tag{2.1}$$

where C_n^2 is known as the structure coefficient (see, for example, Barletti *et al.*, 1977). Under these circumstances the average point-spread function of an aberration-free optical system of 'infinite' aperture is the same as that of a diffraction-limited system of finite diameter (Hufnagel & Stanley, 1964; Fried, 1966). The effective diameter of the system is known as the Fried parameter r_0 and is given by

$$r_0 \propto \lambda^{6/5} I_1^{-3/5}, \tag{2.2}$$

where λ is the wavelength and I_1 is the integrated value of the structure coefficient along the path through the atmosphere

$$I_1 = \int_0^\infty C_n^2(\mathbf{x})\,\mathrm{d}s. \tag{2.3}$$

The scale of the variation of the structure coefficient is found to be large in comparison with the scale of the turbulence itself, so the assumption of isotropic turbulence is not seriously violated.

A fuller analysis, taking account of the finite aperture of the telescope, leads to a more complicated expression for the shape of the average point-spread function, which depends also on the number of speckles present (see Dainty, 1975). But the result is again a function of only the Fried parameter, the effective diameter of the optical system. By comparing photoelectric scans made under different seeing conditions and assuming that the granulation pattern is statistically steady, Ricort & Aime (1979) have sought to determine the set of Fried parameters, one for each scan, and the true rms brightness fluctuation in the photosphere (see Section 2.5.2). One should note that this theory rests on an appeal to statistical averaging over the atmospheric turbulent elements; so only a series of observations can be corrected in the mean. It is not possible to correct an individual observation except by a simultaneous measurement of the wave-front distortion, a feat which is, at present, beyond our capability.

2.3 Properties of the photospheric granules

2.3.1 Shape

It is evident from Figs. 2.1, 2.2. and 2.4(*b*) that many of the granules have very irregular shapes. Polygonal and elongated outlines are very common, and the granules fit neatly together in the pattern like the stones in a crazy-paving path. In other words, in most cases adjacent granules have parallel boundaries on their common sides, the dark channel separating them being of uniform width. The fact that many granules have polygonal outlines was first pointed out by Strebel (1933), whose best photographs show areas in which the resolution is comparable to that of stratospheric photographs such as Fig. 2.1 (see Strebel, 1932: Fig. 1; 1933: Fig. 5). However, Strebel's discovery apparently attracted little attention and as the years passed by was largely forgotten (see Section 1.3). A number of later workers (Macris, 1953; Bray & Loughhead, 1958; Blackwell, Dewhirst & Dollfus, 1959) established that a significant proportion of the granules are non-circular, but their photographs lacked sufficient resolution to reveal the characteristic straight sides of many of the granules, well shown on Fig. 2.4(*b*). In fact, not until the advent of the stratospheric photographs was the true nature of the granulation pattern generally appreciated (e.g. Musman, 1969).

To summarize, the granulation can succinctly be described as an irregular, cellular pattern of bright elements separated by narrow, dark channels. This basic observational fact provides one of the chief reasons for regarding the granulation as a convective phenomenon (see Section 5.2.1).

It should be remarked, though, that the difference of topology between granules and intergranular lanes apparent to the eye is partly physiological in character. The eye tends to effect a division between bright and dark at a contour which lies approximately midway between the maximum and minimum brightness levels of the photograph. At this level, granules appear as features circumscribed by dark lanes. If the brightness level chosen to divide granule from intergranule were to be lowered a few per cent, a different appearance would result (see Karpinsky, 1980a).

2.3.2 Diameter and fractional area

Any theoretical discussion of the photospheric granulation requires a knowledge of some parameter expressing the scale of the pattern. At first sight it might seem that the average granule diameter would be a convenient parameter to use. Unfortunately, direct measurements of the diameters of the individual granules provide only a rough guide to their true dimensions. In fact, the measurement of granule size is actually a complex *photometric* problem involving not only the determination of intensity from the photographic density but also correction for the instrumental profile of the telescope and, in the case of ground-based observations, of the atmosphere. Only when the influence of these factors is accurately known can the true dimensions be inferred.

The measurement is made even more problematical by the failure of various authors to follow any well-defined and objective procedure. Visual inspection of Figs. 2.1 and 2.2 shows that many of the granules appear to have diameters in the range 1-2″ of arc. Accordingly, if one is content with a rough guide to the true granule diameter, a representative, although not necessarily *average*, value - uncorrected for instrumental effects - is $1''.5$ (1100 km). This figure is close to the values given by many workers before photographs of the quality of Fig. 2.1 had been obtained; more recent estimates are collected in Table 2.1. A close examination of Fig. 2.1, though, reveals traces of structure within both the larger granules and the intergranular lanes. Some authors ignore these traces and treat the larger structures as entities (Schröter, 1962; Bray & Loughhead, 1977); others count the fragments and traces as distinct structures (Namba & Diemel, 1969; Karpinsky, 1980a, b). The latter authors thus find systematically smaller values for the granular diameter.

Accordingly, the granule diameter is of little value when we require a parameter for comparison with theory. A more objective parameter is suggested by the topology described in Section 2.3.1: a granule is defined as the area, A_g, over which the intensity is greater than the mean intensity. Since the structures are generally polygonal they have no unique diameter, but an effective diameter may nevertheless be defined as

$$D_g = 2(A_g/\pi)^{1/2}.$$

To our knowledge Karpinsky is the only author to have used such an objective criterion in a quantitative analysis of the solar granulation. He analysed some excellent white-light photographs from the 1970 flight of the SSSO, but his results have so far appeared only in abbreviated form. In his investigation the reference level was defined some 4% below that of the mean brightness, which forced the granules and the dark intergranular patches to have the same

Table 2.1. *Mean diameter of the photospheric granules*

Reference		Value
Bray & Loughhead (1958)		$1''.4$
Macris (1959)		$1''.3$
Michard (1961)		$1''.3$
Schröter (1962)		$1''.25$
Harvey & Ramsey (1963)		$1''.3$
LaBonte *et al.* (1975)		$1''.5$
Bray & Loughhead (1977)		$1''.35$
	Average	$1''.34 = 970$ km
Namba & Diemel (1969)		$1''.1$ (mean) $1''.2$ (rms)
Karpinsky (1980b)		$0''.96$ (mean) $1''.11$ (rms)

topology. This choice resulted in the granules having greater total area and greater individual areas, as well as larger brightness excursions from the reference level, than the dark patches. Moreover, his figures exclude 'small' granules and dark patches. Because of the selection effects present in both this investigation and that of Namba & Diemel (1969), their respective results will be quoted separately and will not be included in any average. Thus the modern measurements of granule diameters, despite varying procedures, are quite consistent, as may be seen from Table 2.1. The mean value is 1″.34 (970 km).

There is a large dispersion in granule size: the smallest granules visible on Fig. 2.1 which are nevertheless fully distinguishable - i.e. completely surrounded by dark material - have an apparent diameter of about 0″.4 (300 km), a figure comparable with the resolving limit of the telescope used; the largest granules, on the other hand, have a diameter (largest dimension) of about 3″.2 (2300 km). That the distribution of diameters has a long tail towards large values is shown by the fact that the root mean square diameter is some 10% greater than the mean diameter (Namba & Diemel, 1969).

The size distribution of the solar granulation is extremely uniform over the solar surface; only close to sunspots is a systematic shrinkage of the mean granule diameter apparent. Estimates of granule areas by Schröter (1962) led to an average diameter of 1″.2-1″.3 for granules in the undisturbed photosphere and 1″.0 for granules in the immediate vicinity of sunspots. Macris (1978) found a correlation between the magnetic field strength of the spot and the reduction of the size of the granules within 1-2″ of the penumbral boundary (see also Section 2.6).

Measurements of granule areas used to determine diameters also yield the total area covered by the granules, ΣA_g. If A_T is the total area sampled, the fractional area of granules is

$$\phi_g = \Sigma A_g / A_T.$$

Recent estimates of ϕ_g give an average value of 0.49 (see Table 2.2), implying little or no difference between the fraction of the solar surface covered by granules and that covered by the intergranular lanes.

2.3.3 *'Cell size' and total number of granules on the Sun*

In view of the difficulty of measuring granule diameters, Bray & Loughhead (1959) and Rösch (1959) independently introduced instead the *mean cell size* of the pattern, defined as the average distance between centres of adjacent granules, as a convenient quantitative parameter for characterizing the scale of the pattern. This parameter is independent of photographic contrast and instrumental resolution, provided the individual granules are actually resolved. As we shall see in Chapters 3 and 4, the mean cell size plays an important role in theoretical discussions of convection.

The values for the average distance between granules obtained by direct measurement are collected in Table 2.3. The results of Bray & Loughhead (1977) are shown in Fig. 2.10 which gives the distribution of distances between the centres of contiguous granules, derived from measurements of a group of 83 granules on two good-quality photographs taken with the 30-cm refractor of the CSIRO Solar Observatory, Culgoora. It is evident that the granulation pattern has a well-defined and rather narrow distribution of intergranular distances: in fact 88% of the values lie between 1″.0 and 3″.0 of arc. The mean value derived from these data is 1″.9. The tail extending out to 4″.6 reflects the presence in the pattern of the occasional dark elements mentioned in Section 2.1.

Namba & Diemel (1969) have criticized this procedure for its tendency to smooth out the actual distribution of cell sizes. This can be seen by imagining a square grid with alternately large and small granules at the grid points; in this case the ratio of granule diameters can be made arbitrarily large but the intercell distance remains constant. Moreover, in reality all granules do not have the same number of neighbours so different granules will appear in the statistics with

Table 2.2. *Fractional area of granules*

Reference		Value (%)
Rösch (1959)		56.5
Pravdjuk *et al.* (1974)		44.5
Keil (1977)		47
Wittmann & Mehltretter (1977)		47.6
	Average	49
Namba & Diemel (1969)		51–66
Karpinsky (1980b)		⩾30

Table 2.3. *Average distance between centres of adjacent granules*

Reference	Value
Bray & Loughhead (1959)	2″.9
Rösch (1959)	2″.0
Macris & Banos (1961)	2″.0–2″.9
Rösch (1962)	2″.55
Bray & Loughhead (1977)	1″.94

different weight. Namba & Diemel preferred to draw a line along the intergranular lanes to define each cell and then to measure directly the cell area A_c. The effective diameter, assuming the cells to be circular, is a characteristic cell dimension

$$d_c = 2(\bar{A}_c/\pi)^{1/2},$$

which is analogous to the intercell distance. Namba & Diemel find a value of 1″.5 for this cell dimension, which is significantly smaller than the previous values quoted but is attributable to the inclusion of granular fragments in the counts. Their procedure yields a result equivalent, within a small numerical factor, to simply counting the number of granules N_g within a certain area and calculating

$$d_c = (A_T/N_g)^{1/2}.$$

Table 2.4 lists the values of the surface density of granules, normalized to a standard area of 10″ x 10″, published by various workers. The mean granule

Fig. 2.10. Distribution of the distances between the centres of 83 adjacent granules. Note the rather narrow width of the distribution, 88% of the values lying between 1″.0 and 3″.0 of arc. The mean cell size derived from these data is 1″.9 (Bray & Loughhead, 1977).

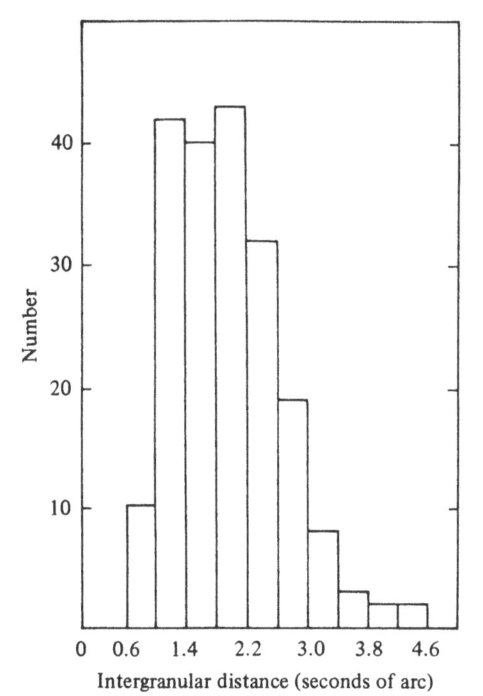

count of 31.5 implies a mean cell area $\bar{A}_c$ of 1.67×10^6 km^2. The corresponding total number of granules on the surface of the Sun is 3.7×10^6, and the characteristic cell dimension is 1″.8 (1300 km).

This dimension is also the distance between the centre of the cell and the centres of its *nearest* contiguous neighbours. Some authors prefer to calculate a characteristic dimension assuming a hexagonal planform; this dimension is then the distance between the centre of the cell and the centres of *all* its contiguous neighbours. Granules, as described in Section 2.3.1, are irregular polygons but their spatial relationships demonstrate a quasi-hexagonality (Hejna, 1980). According to Aime, Martin, Grec & Roddier (1979), the power spectrum of brightness fluctuations can be interpreted in terms of a pattern in which granules are centred at points randomly displaced from a hexagonal grid. The standard deviation of the displacement is only about one-third of the distance between the cells, so the pattern retains noticeable regularity. Interpreted as hexagons, the characteristic cell dimension is 8% larger than that given above, and is thus 1″.9 (1400 km) on average. This value is in good agreement with the only recent direct determination of the average distance between granules, 1″.94 (Bray & Loughhead, 1977).

We may conclude that, despite the differences in methods of measurement, the characteristic scale, or mean cell size, of the granulation is a well-defined quantity; the best estimate at present available is 1″.9 of arc, i.e. 1400 km, the uncertainty in the value being of the order of a tenth of a second of arc.

Macris and his coworkers have reported that the total number of granules on

Table 2.4. *Surface density of granules and characteristic cell size*

Reference	Surface density per 10″ × 10″ square	Cell size[a]
Rösch (1959)	30	1″.8
Schröter (1962)	31.9	1″.77
Birkle (1967)	32.6	1″.75
Pravdjuk *et al.* (1974)	30.0	1″.82
LaBonte *et al.* (1975)	29.5	1″.84
Muller (1977)	33.0	1″.74
Mehltretter (1978)	33.5	1″.73
Kawaguchi (1980)	31.2	1″.79
Average	31.5	1″.78 = 1300 km
Namba & Diemel (1969)	51.3–58.0	1″.4–1″.3
Karpinsky (1980b)	31	1″.8

[a]Calculated on the assumption of a square shape.

the Sun (Macris & Elias, 1955), or equivalently the mean cell size (Macris & Banos, 1961), is dependent on the phase of the solar cycle. This dependence was not confirmed by Birkle (1967), using a homogeneous set of observations from Potsdam extending over 15 years. In view of the errors inherent in the determination of these parameters in the absence of uniform material of 'stratospheric' quality, we conclude that it has yet to be established that any of the properties of the photospheric granulation show a dependence on the phase of the solar cycle. However, in Section 5.5.2 we shall find indirect evidence for a possible dependence of granulation properties on activity and solar latitude.

2.3.4 *Granular contrast*

The granular contrast may be defined as the quantity

$$C = \frac{I_{max} - I_{min}}{\frac{1}{2}(I_{max} + I_{min})},$$

where I_{max} and I_{min} represent the brightness of the granules and the intergranular material respectively. This quantity is the most direct expression of the difference in the thermodynamic parameters, most particularly temperature, between the granules and intergranular lanes. It is unfortunate, therefore, that the measurement of the contrast is attended by difficulties similar to those encountered in diameter measurements (Section 2.3.2). Apart from an ambiguity in allocating intergranular lanes to particular granules there is also a photometric problem due to the usually inadequate resolution of the photographs.

The problem of determining the granular contrast is almost exactly analogous to the familiar problem of measuring the true central intensity of a Fraunhofer line. This is possible only when the width of the instrumental profile of the spectrograph is substantially less than the width of the line. Similarly, in measuring the granular contrast, the instrumental profile of the telescope ideally should be substantially narrower than the width not only of the granules themselves but also of the *dark lanes between them*. A careful scrutiny of Fig. 2.1 shows that even on this photograph - taken with a 30-cm telescope above the Earth's atmosphere - the apparent width of the dark lanes is often less than 0″.5 of arc. The true width must be even smaller since the effect of the finite resolution is to widen the dark lanes. It follows that it must be comparable with or smaller than the theoretical resolving limit of a 30-cm telescope (0″.4).

In fact, in order to obtain a truly reliable value for the granular contrast the following conditions must first be satisfied:

(1) the effective resolving limit of the telescope should be several times smaller than the true width of the dark lanes (a 120-cm telescope would probably be adequate);

(2) the instrumental and (if ground-based) the atmospheric point-spread functions need to be accurately known.

Such stringent conditions have yet to be met fully, so it is not surprising that the published values of the granular contrast listed in Table 2.5 show considerable scatter, part of which however may be due to a variation with wavelength (see Section 2.5.3). This table includes only direct measurements of contrast; values derived from rms brightness fluctuations are excluded. Probably, Bray & Loughhead's (1977) value of 21-7% at 550 nm - carefully corrected - remains the most reliable determination to date.

The average distribution of brightness across a granule and an intergranular lane has attracted little attention from observers. Information regarding it is available only indirectly in the form of probability density histograms which give the relative area of the solar surface as a function of brightness. Several authors (Edmonds, 1962b; Pravdjuk, Karpinsky & Andreiko 1974; Keil, 1977; Wittmann, 1981) have published brightness histograms, all of which show appreciable deviations from a normal distribution. They are skewed toward large positive fluctuations; very bright granules are commoner than very dark intergranular regions. The histograms are also more flattened than a normal distribution, i.e. values near the mean brightness are less common than would be expected in a random intensity pattern. These two properties are described mathematically by the skewness S and excess E of the distributions,

$$S = m_3/m_2^{3/2}$$

$$E = (m_4/m_2^2) - 3,$$

Table 2.5. *Contrast of the photospheric granulation*

Reference	Wavelength (nm)	Aperture (cm)	Measured contrast (%)	Corrected contrast (%)
Blackwell *et al.* (1959) (corrected by Gaustad & Schwarzschild, 1960)	530	29	10	16
Rösch (1959)	595	23	11.7	23
Rösch (1962)	600		15	22
	461		13	20
Krat *et al.* (1972)	465	50	16–35	
Bray & Loughhead (1977)	550	30	14.7	21–7
Karpinsky (1980b)	465	50	20	36
Alissandrakis *et al.* (1982)	520	38, 50	16.5	18–25

where the moments m_i are

$$m_i = \Sigma(I - \bar{I})^i.$$

Both the skewness and excess are zero for a normal distribution. Table 2.6 compares measurements of the skewness. Apart from Edmonds (1962b) no-one seems to have recognised the diagnostic value of these distributions (see Section 3.4.5).

2.3.5 *Diversity in size and brightness*

Although the majority of the granules appear to have diameters in the range 1-2″ of arc (725-1450 km), it is also evident that some have diameters of less than half a second of arc, while others exceed several seconds (see Figs. 2.1 and 2.2).

Good granulation photographs often show isolated regions where smaller granules seem to predominate (Blackwell *et al.*, 1959; Edmonds, 1960). However, these appear to be chance groupings which gradually disappear as new granules form and others decay.

The individual granules also show a considerable diversity in their brightness. The brightness differences are perhaps somewhat better shown in Fig. 2.3(*a*) than in Fig. 2.1. Among granules of comparable size it is easy to find some which are distinctly brighter than average and others distinctly fainter than average.

It is of interest to enquire what correlation, if any, exists between the size and brightness of the granules. This question was first studied by Bray & Loughhead (1958) in the course of their investigation into changes of the shape, size and brightness of the granules during their observed lifetimes (Section 2.3.7). As a consequence of the limited resolving power, it was possible to classify the granules in a rather crude and qualitative way only. They found that bright granules were about one-half as numerous as those of medium brightness and three times as numerous as the faint ones. Particularly note-

Table 2.6. *Skewness and excess of granular brightness distribution at disk centre*

Reference	Skewness	Excess
Edmonds (1962b)	0.308, 0.126	
Pravdjuk *et al.* (1974)	0.41	−0.66
Wittmann (unpublished)	0.30	−0.37

worthy was the almost complete absence of large, faint granules. In the case of small granules, bright granules were very rare and less numerous than faint ones.

These findings are generally confirmed by stratospheric photographs taken with telescopes having more than twice the resolving power (e.g. Namba & Diemel, 1969). Mehltretter (1978) distinguished four types of granule with the following frequencies: small fragments (52%), average-sized fragments (28%), large and compact granules (16%) and ring-like exploding granules (4%) (see also Section 2.3.7). These Spektrostratoskop results are given in Fig. 2.11, which shows the scatter diagram of the mean granular brightness versus the granular area measured at a level 5% brighter than the mean. These data have not been corrected for the instrumental profile.

It is possible that the very weak correlation between brightness and size suggested by these results is simply a consequence of the greater radiative cooling time to be expected for the larger granules. If so, the diversity in brightness is directly related to the diversity in size. On the other hand, the diversity in size (and shape) itself can have only a hydrodynamic explanation. Many authors have suggested that the irregular appearance of the granulation reflects the fact that it represents a convective process well beyond the state of marginal instability. A detailed discussion will be found in Section 5.3.

Fig. 2.11. The mean brightness of solar granules as a function of area, uncorrected for instrumental blurring. For very small granules the apparent correlation between brightness and area is due to finite resolution. The mean granule area in this sample corresponds to a diameter of 1$''$.24 of arc (Mehltretter, unpublished).

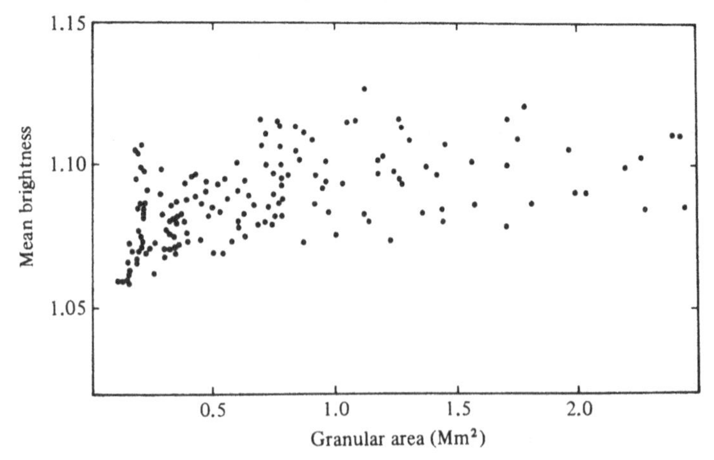

2.3.6 *Lifetime*

To obtain an accurate value for the lifetime of the photospheric granulation it is necessary to analyse carefully a sequence of good-quality photographs of the same region, extending over a period of at least twenty minutes and preferably longer. The individual existence of a number of selected granules must be followed from one photograph to another in the sequence. Although very tedious to apply, this method gives directly the quantity required. Moreover, with a sufficiently long sequence, it allows a determination not only of the average or most probable lifetime but also of the *distribution* of lifetimes. Finally, with photographs of adequate quality it would enable one to determine what correlation, if any, exists between the lifetimes of individual granules and their other physical characteristics, e.g. brightness, size and shape. It suffers from the disadvantage, however, that if the photographs are of poor, or uneven, quality rather subjective decisions may sometimes have to be made as to the existence or non-existence of particular granules at particular times.

The first adequate determination was made by Macris (1953) using a 22-min sequence obtained by Lyot in 1943. Macris found a value of 6–8 min for the most probable lifetime although individual values as high as 15–16 min were recorded. Bray & Loughhead (1958) obtained results in broad agreement with those of Macris. However, owing to the relatively short duration of the sequence (10 min) the starting and ending times of many of the granules fell outside the period of observation, suggesting that the most probable lifetime is somewhat greater than the value of 6–9 min actually found. In agreement with this conclusion Rösch & Hugon (1959) reported that many granules persist for about 10 min, whilst Rösch (1962) recorded lifetimes as long as 20 min.

A 48-min sequence obtained during the Spektrostratoskop flight was studied by Mehltretter (1978). He selected seven areas each of $5'' \times 43''$, one of which is shown in Fig. 2.12. The average interval between frames was 2 min. Tracing granules forwards and backwards in time from a 'master' frame, Mehltretter found that the number of granules surviving at time $t, N_g(t)$, was related to the number present on the master $N_g(0)$ by the exponential law

$$N_g(t) = N_g(0) \exp(-|t|/t_g)$$

where the decay time t_g was 8.1 min for the total granule sample. This quantity is the average time that a granule can be traced *both backwards and forwards* from $t = 0$; the mean lifetime is twice as large, 16 min. According to Mehltretter, this value is typical of granules which fragment (see Section 2.3.7). Granules which subsequently merge have shorter lifetimes, those which dissolve have longer lifetimes. Very similar results were obtained earlier by Simon (1967) using an exceptional sequence obtained at the Sacramento Peak Observatory. He also derived a decay time of 8.1 min.

Modern values of granular lifetimes are collected in Table 2.7.

2.3.7 *Evolution*

A much more difficult observational problem than the determination of granule lifetimes is the study of their evolution. This question, however, is of great interest for its bearing on the underlying hydrodynamic processes. Time-dependent theoretical models of the granulation have only recently begun to be investigated (see Section 5.2.4); an exact knowledge of the manner in which granules evolve and how it varies, e.g. with granule size, could throw useful light on the physics of the non-linear interactions believed to occur.

Table 2.7. *Lifetime of photospheric granules*

Reference	Lifetime (min)
Macris (1953)	6–8
Bray & Loughhead (1958)	>7–8
Rösch & Hugon (1959)	10
Macris & Prokakis (1963)	8.2 (median)
Simon (1967)	16 (mean)
LaBonte *et al.* (1975)	~15 (mean)
Mehltretter (1978)	16 (mean)

Fig. 2.12. Time sequence illustrating three characteristic modes of granule decay: dissolution, merging, and fragmentation. The photographs, ~2 min apart, were obtained with a 30-cm balloon-borne telescope on 17 May 1975; each frame measures 5″ × 43″ of arc. (By courtesy of J. P. Mehltretter.)

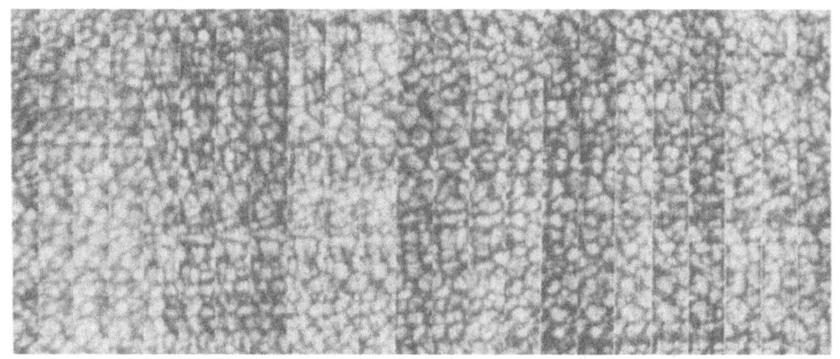

The greater difficulty is due to the fact that the *description* of the individual granules demands photographs of much better quality than those required for their mere *identification*. The first systematic attempt to detect changes in the brightness, size, and shape of the individual granules during their observed lifetimes was made by Bray & Loughhead (1958). Using the 10-min sequence of photographs from which the lifetime was measured, 140 granules were first classified in regard to brightness, size, and shape according to their appearance on a very good 'master' photograph near the middle of the sequence. The same granules were then described according to their appearance on each of a number of other photographs occurring before and after the master, thus enabling the development of the individual granules to be followed from photograph to photograph over an average period of nearly 7 min.

The overall impression was that of stability. There was some tendency among granules showing change for size increases to predominate over decreases, but brightness increases and decreases occurred with equal frequency.

Examples of the types of change observed in individual granules were also given by Rösch & Hugon (1959), but they failed to reach any definite conclusions about the general mode of evolution of the granules. In a later note Rösch (1962) gave the first substantially correct description of the evolution of a typical granule, based on Pic-du-Midi observations. He asserted that, after a granule is formed, its diameter begins to increase, in general, until it reaches about 2″ of arc. Then it breaks up into several small granules which fade and vanish at the place where they appeared.

This picture has since been confirmed by studies of time series of stratospheric quality. Mehltretter (1978), for example, employed a 48-min sequence of white-light photographs taken during the Spektrostratoskop flight already mentioned in Section 2.3.6. The average time separation of the frames was 2 min. Mehltretter concentrated attention on seven fixed areas of 5″ x 43″ (one of which is shown in Fig. 2.12) and traced the history of all granules within the areas forwards and backwards in time. He summarized the results by distinguishing two time scales. On a time scale of less than 6 min the granular evolution is predominantly that of expansion of either single granules or groups of granules. The apparent expansion proceeds at a rate of 0.8 km s^{-1} on average whilst asymmetric displacements achieve peak rates of 2–3 km s^{-1}. Over longer time scales the individual granules either fragment or drift. The larger and better-defined expansion centres evolve more or less coherently and absorb adjacent fragments of earlier origin during their expansion. There appeared to be no general correlation between size and lifetime; even small fragments can be long-lived, although they may undergo considerable changes of size, shape, and brightness. This conclusion has been contested by Kawaguchi (1980), who finds a good correlation between lifetime and maximum size for those granules which simply fade away and do not merge or fragment.

Some initially bright granules continue to expand rapidly and more or less symmetrically up to diameters of 3-5″ before fragmenting. Fragmentation is presaged by the formation of a dark central spot which gives the granule a ring-like appearance. The spot then develops dark radial spokes that fragment the structure. This phenomenon has been described as an 'exploding' granule; it was discovered by Carlier, Chauveau, Hugon & Rösch (1968), and has since been studied by Allen & Musman (1973) and Namba & van Rijsbergen (1977). Mehltretter found that this behaviour is just a short-lived and extreme example of the normal granular evolution by expansion and fragmentation. This interpretation has been corroborated by Kitai & Kawaguchi (1979), who obtained at the Pic-du-Midi Observatory a long sequence (46 min) of very high-quality photographs.

Their analysis of a 4.5-min portion, with intervals between photographs of 0.5 to 1 min, suggested that dark spots occur with a frequency of about 3.5 per 10″ × 10″ square, and have a lifetime of about 2 min. Since they are ten times less frequent than granules themselves but have a lifetime almost a factor of ten shorter, there is a strong implication that many, if not most, granules develop a dark spot during their lifetime. Most dark spots evolve into a dark notch, connected to the intergranular boundary, and then into a new lane fragmenting the granule after a minute or so.

This appears to be the commonest mode by which a granule loses its identity. Mehltretter reported that 50% of all granules split, producing from 2 to 5 fragments. The mean number of fragments is given as 2.8 by Mehltretter and 2.3 by LaBonte, Simon & Dunn (1975). Other granules disappear by simply fading into the intergranular background or by being absorbed by a neighbouring granule. The apparent frequencies of these two processes seem to depend on individual judgement, or seeing quality, or, more likely, both: Mehltretter describes 20% of granules as fading away and 30% as merging, whilst LaBonte *et al.* give figures of 60% and 4% for these two processes.

Kawaguchi (1980) has provided some clarification by a further analysis of the Pic-du-Midi material. He observes that small granules tend to fade away, and that almost all large granules fragment or merge. 50% of his sample achieved a maximum diameter of 1″.5, and of these 45% faded away. 40% reached a maximum diameter of between 1″.5 and 2″.5, and 10% of these disappeared by fading. The remaining 10% of the sample, with maximum diameter greater than 2″.5, all merged or fragmented.

All authors agree that granules are born, almost without exception, from previous granular fragments. Occasionally, a granule appears to develop out of a faint patch of brightening in an extensive area of intergranular material, but most granules represent a resurgence of an earlier manifestation. This view has been elaborated by Kawaguchi (1980) who showed that repeated fragmentation and regeneration lead to 'families' of granules with a typical diameter of 3-5″. The length of time that a family persists is related to both the maximum size of the largest member and to the total number of members. Some 30% of the families traced in an area of 17″ x 17″ still existed at the end of the 46 min covered by Kawaguchi's sequence.

The recognition of such families of granules, if confirmed (compare the description of 'mesogranulation' given in Section 2.7), is a potentially significant development. It means that the scales and lifetimes of the individual granular structures are much less than those of the underlying convective process that produces them. Such correlations extending in both space and time are admitted in neither the picture of steady cellular convection nor of turbulent mixing-length convection but seem to be discernible in the more realistic time-dependent models of Section 5.2.4.

2.4 Extension of granules into the upper photosphere

2.4.1 Granulation near the extreme solar limb

The visibility of the photospheric granulation decreases towards the limb until, finally, no trace of the pattern can be detected. A determination of the distance from the limb at which it finally disappears provides an estimate of the height to which the granules extend into the upper photosphere (Plaskett, 1955; de Jager, 1959: see p. 83). This question is of the greatest importance since, for a variety of reasons, we need to know the structure of the inhomogeneous photosphere not only in the deeper layers accessible to observation in continuous radiation but also in the higher layers where the Fraunhofer lines originate.

Rösch (1957) found that on good photographs the granulation remains visible to within less than 10″ of arc from the limb, and sometimes to less than 5″. Although Edmonds (1960) concluded on the basis of an examination of stratospheric photographs that the granulation disappears at 33-21″, Loughhead & Bray (1960) were able to confirm Rösch's value. Moreover, in a subsequent paper Edmonds (1962b) revised his estimate to 15-10″. The various estimates are collected in Table 2.8.

Fig. 2.13, taken from Loughhead & Bray (1960), clearly demonstrates that the granulation remains visible very close to the limb. The photographs show overlapping regions of the Sun in the neighbourhood of the west limb; both enlargements were made from the same original negative, using intermediate negatives of slightly different densities in order partially to compensate for

Table 2.8. *Distance from the limb at which the granulation disappears*

Reference	Heliocentric angle	Estimated distance (sec of arc)
Rösch (1957)	82°–84°	10–5″
Edmonds (1960)	75°–78°	33–21″
Loughhead & Bray (1960)	82°–85°	10–4″
Edmonds (1962b)	80°–82°	15–10″
Muller (1977)	86°	<2″

Fig. 2.13. Granulation near the extreme solar limb, photographed with the Sydney 13-cm photoheliograph (Loughhead & Bray, 1960). Both prints were made from the same original negative; the white line indicates the position of the actual limb, derived from the negative. The upper scale gives the heliocentric angle, the lower scale the distance from the limb in seconds of arc.

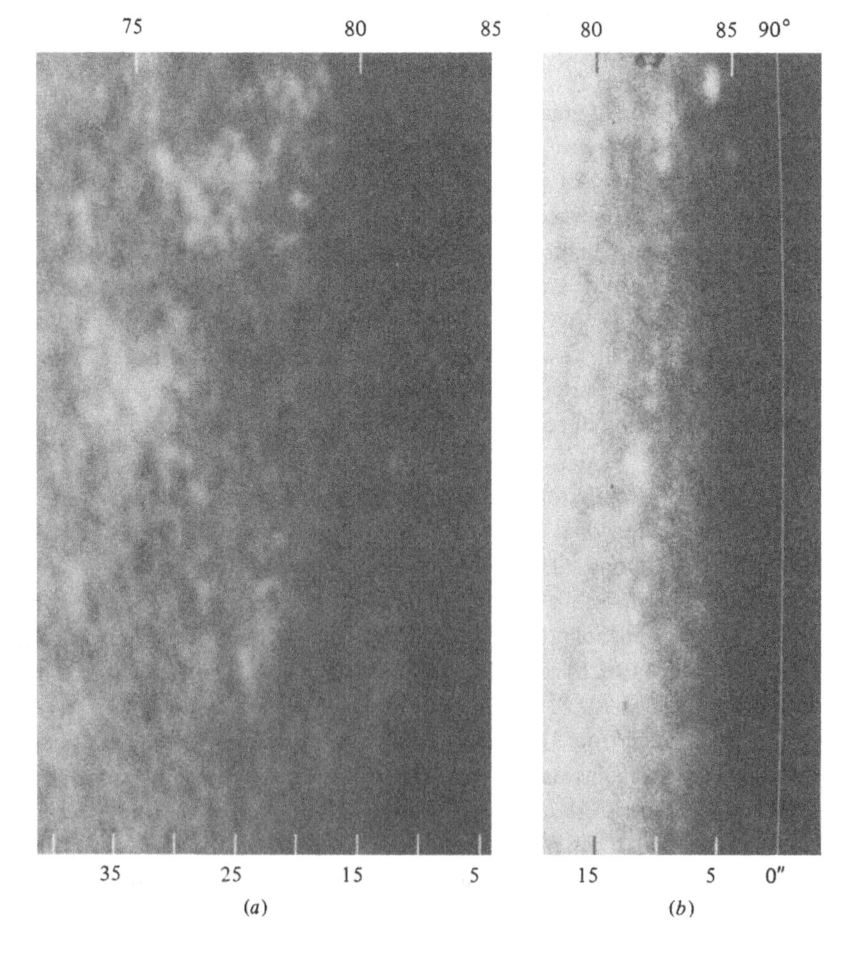

limb darkening. The white line on Fig. 2.13(*b*) indicates the position of the actual limb, derived from the original negative.

The brighter features are facular granules - long-lived, small-scale (1-2″) brightenings commonly, but not exclusively, seen in the neighbourhood of sunspots and thought to be associated with strong magnetic fields (Bray & Loughhead, 1967: Section 3.3; Muller, 1977). Apart from these, a number of granules can be seen less than 10″ from the limb; in fact, one rather bright granule can be seen only 4″ from the limb. Even in regions where individual granules are hard to distinguish the photograph gives the impression of a low contrast foreshortened picture of the ordinary granulation. No granules are visible on the original negative in the last 4″ to the limb (this region does not appear in Fig. 2.13(*b*)). On the basis of Pic-du-Midi photographs with an apparent resolution of 0″.3 Muller (1977) claims that the normal granulation is always visible up to 2″ from the limb (heliocentric angle = 86°). Even more remarkably the SSSO observations reveal fine structure *within* 2″ of the limb whose contrast is almost independent of the distance from the limb (Pravdjuk *et al.*, 1974).

The interpretation of observations close to the solar limb is by no means straightforward. The factors which need to be considered are the finite resolution of the telescope, the high degree of foreshortening, and the three-dimensional structure of the individual granules. The question of the visibility of the granulation near the limb is not merely a geometrical problem, but one involving the theory of radiative transfer in a non-uniform medium (see Sections 4.4 and 5.2.2).

Let us take it as established that the granules are visible up to 5″ from the limb (heliocentric angle = 84°); this implies that the convective pattern can be followed up to a nominal optical depth of $\tau = \cos 84° = 0.1$. The granulation is thus seen to extend far beyond the level, $\tau \sim 1$, above which the atmosphere becomes convectively stable according to the Schwarzschild criterion (see Section 3.3). Since the weaker Fraunhofer lines and all but the inner cores of the stronger lines (roughly, all points in the profile at which the intensity is greater than 30% of the continuum value) are formed at or below $\tau = 0.1$, we should expect to find evidence of the presence of the granulation in observations of the Fraunhofer lines.

Fig. 2.14(*a*) shows a white-light picture (taken through a 10 nm filter centred at λ575 nm) of the granulation obtained by G. Ceppatelli and R. Muller at the Pic-du-Midi Observatory, while Fig. 2.14(*b*) shows one taken 1.5 min later using a 0.8 nm filter whose passband admitted the strong Fe I line at λ430.8 nm. The presence of lines as well as continuum in the passband raises the mean level from which the radiation arises (see Fig. 2.15). A rough estimate is provided by the observation that the mean intensity in the passband is about 50% of the con-

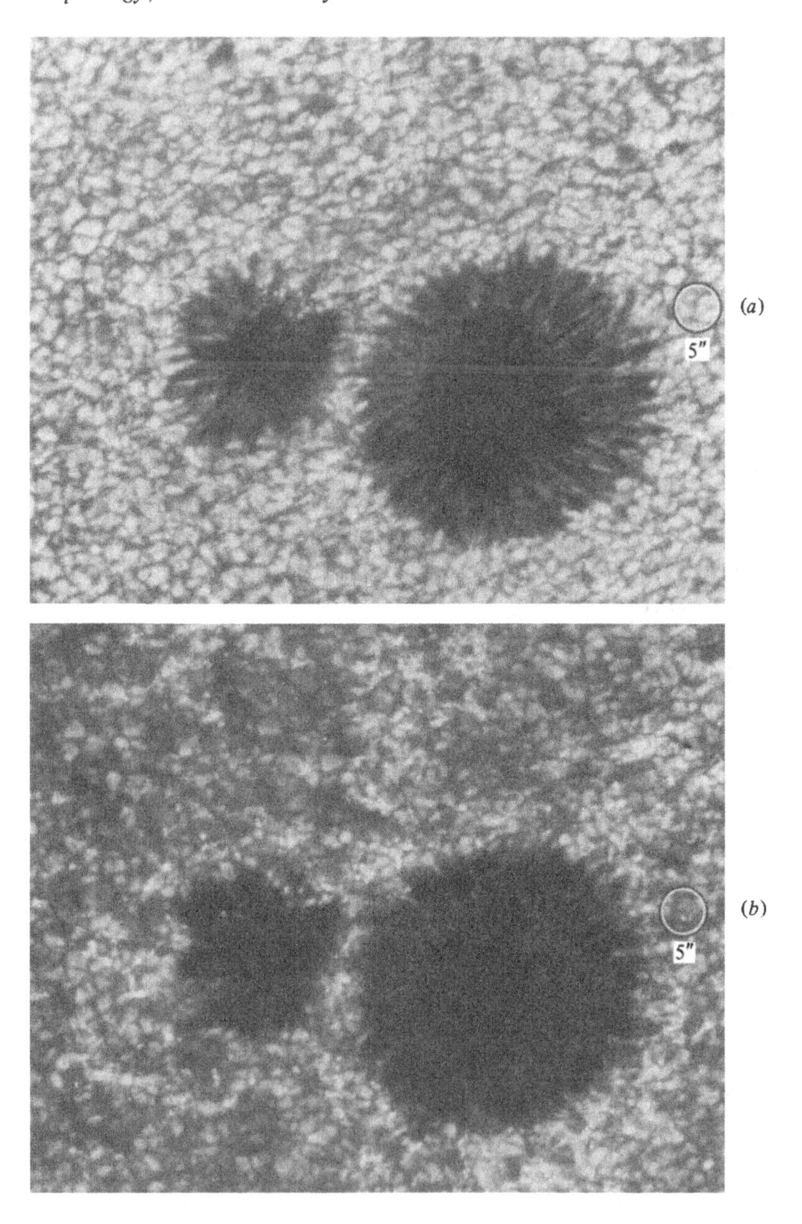

Fig. 2.14. Photospheric granulation near sunspots photographed on 4 June 1980 with the 50-cm refractor of the Pic-du-Midi Observatory through
(*a*) a 10 nm filter centred at λ575 nm;
(*b*) a 0.8 nm filter centred at λ430.8 nm.
The 'continuum' photograph (*a*) shows the closer packing of the granules in the immediate vicinity of the spots. Stretching away from

tinuum value; this implies a mean level of formation of $\tau = 0.2$ according to the Eddington–Barbier relation (see Section 4.4.1), i.e. about 100 km above $\tau = 1$. This level corresponds to viewing the photosphere at a heliocentric angle of 78°, or 20″ from the limb. Fig. 2.14(*b*) again shows clearly the facular granules, which form a very broken network with a scale of some tens of seconds of arc. In the background, though, the granulation pattern, which dominates the white-light photograph (Fig. 2.14(*a*)), shows up quite clearly.

Such 'narrow-band' photographs are not themselves conclusive evidence that the granulation pattern extends into the middle photosphere since some lower photospheric light is admitted by the filter and can form a weak parasitic image. They are nonetheless fully consistent with the other evidence that the granulation can be detected in radiation arising nominally at heights of 100 km or more. We once again remind the reader that the interpretation of such observations in terms of the distribution of thermodynamic quantities associated with the granulation involves the theory of radiative transfer in a non-uniform medium (see Sections 4.4 and 5.2.2).

2.4.2 *Granule velocities*

Fundamental to a detailed understanding of the physics of solar convection is a knowledge of the magnitude of the convective velocities in the photosphere. In principle, the upward and downward velocities of granules and intergranular material respectively can be directly measured by means of the Doppler shift of Fraunhofer line radiation emitted by the moving gas.

In using this method we must remember that the spectral lines employed in velocity measurements originate at higher levels in the photosphere than the continuous radiation that provides direct photographs of the granulation in the central region of the disk. Fig. 2.15 is a reproduction of a diagram prepared by Edmonds (1962a) showing the contribution curves for a number of lines typical of those used in measuring photospheric velocities and, in addition, the contribution curve for the continuum at λ500 nm. One can see that the separation between the peaks of the line contribution curves and that of the continuum curve is of the order of 100–200 km. In every case, the amount of overlap between the curves is relatively small. (The effective heights to which velocity measurements refer are more fully discussed in Section 4.4.1.) On the other hand,

the spots are tongues of abnormal diffuse granulation, which generally coincide with the regions of bright facular granules in the upper photospheric photograph (*b*). Elsewhere, the granulation is very similar in both upper and lower photosphere. (By courtesy of G. Ceppatelli & R. Muller.)

as we have seen in Section 2.4.1, the fact that the granulation pattern remains visible up to 5″ of arc from the limb (and possibly even closer) demonstrates that individual granules must persist as coherent visible structures at least up to a nominal optical depth of 0.1. There is no *a priori* reason, therefore, why the observed small-scale photospheric velocities should not, at least in part, reflect the actual motions of the granules.

The first spectra having a spatial resolution sufficient to resolve individual photospheric granules were obtained by R. S. Richardson in 1949, using the spectrograph of the 46-m Mt Wilson solar tower equipped with a new Babcock grating. These spectra showed prominent bright and dark streaks running parallel to the dispersion, which were produced by the granules and the intergranular spaces falling on the spectrograph slit. The streaks themselves were intersected by the solar absorption lines in a definite zigzag manner, indicating the presence of both upward and downward velocities at the points of intersection.

The facts are well illustrated in Fig. 2.16(*a*) which shows part of a very good-quality spectrogram obtained at the Kitt Peak National Observatory under

Fig. 2.15. Contribution curves computed by Edmonds (1962a) for the continuum (λ500 nm) and for a number of spectral lines typical of those used in measuring small-scale photospheric velocities. The height separation between the peaks of the line contribution curves and the maximum of the continuum curve is of the order of 100–200 km.

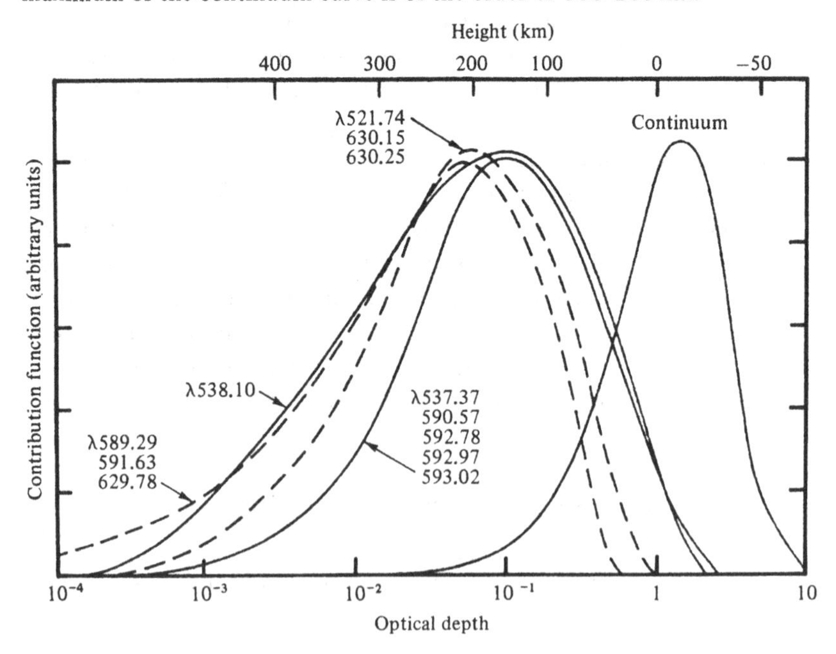

Fig. 2.16. High-resolution granulation spectrograms obtained at (*a*) Kitt Peak National Observatory, showing detail in the continuum. Thirteen prominent granules are present between *A* and *B*, a distance of 34″. Each narrow intergranular lane is marked by a red displacement of the lines, indicating a downward velocity. (By courtesy of Association of Universities for Research in Astronomy, Inc., Kitt Peak National Observatory.)
(*b*) Pamir Observatory, showing detail in the lines. The sharp displacements of the lines associated with the granulation can be traced even in the core of the strongest line present. (By courtesy of V. N. Karpinsky, Pulkovo Observatory.)

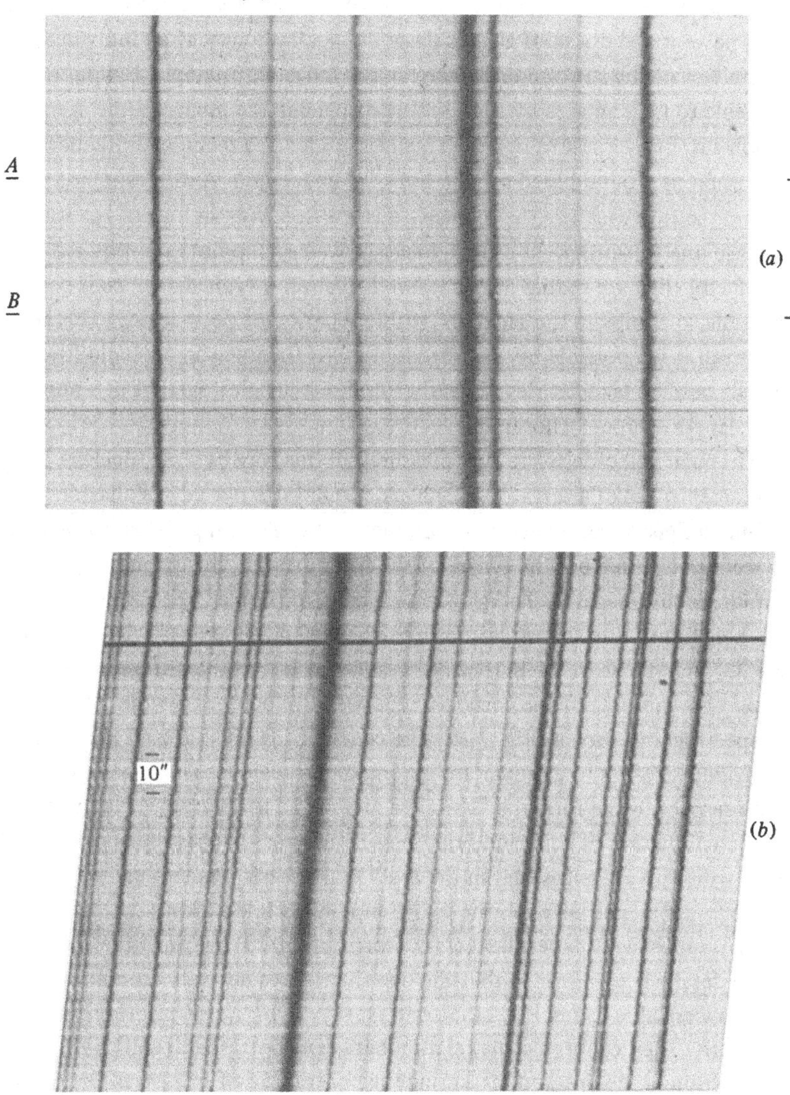

conditions of excellent seeing and analysed by Kirk & Livingston (1968). The photograph has been printed to bring up the detail in the continuum. The four most prominent Fraunhofer lines visible are those of Ba II λ585.37, Ca I λ585.75, Ni I λ585.78 and Fe I λ585.96 nm whose equivalent widths are 0.0055, 0.0132, 0.0056 and 0.0074 nm respectively. Between the points A and B, 34″ apart, thirteen bright granular patches of various sizes can be discerned. The associated Doppler shifts to the violet vary from granule to granule but each one is clearly bordered by a narrow, dark intergranular streak possessing a prominent redshift.

A spectrogram is, however, only a one-dimensional sampling of the granulation pattern and does not always show the peak velocity of an individual granule or of the surrounding intergranular lanes. In principle, it would be possible to map an area by successively displacing the image on the spectrograph slit, but variations in seeing quality render this technique practically useless as far as ground-based observations are concerned. A different, and better, technique is to measure the intensity distribution over an area of the solar surface one wavelength at a time. Fewer wavelength points are required to derive a velocity than are spatial scans required to cover a typical area, so it becomes possible to complete the sequence within a period of good seeing. This approach has been made possible by the introduction of tunable narrow-band filters which can be used to take a series of nearly simultaneous photographs at a number of wavelengths across a chosen line.

Granular velocities were first measured in this way by J. M. Beckers of the Sacramento Peak Observatory (Beckers, 1968a). He placed a Wollaston prism behind a Zeiss 0.025 nm filter to produce two 0.025 nm pass bands, each displaced 0.012 nm from the centre of the filter profile. This arrangement created simultaneous images in the light of each wing of the Fe I λ656.92 nm line. Beckers found that the variations in brightness in the red wing I_r were much greater than those in the blue wing I_b, and noted that the difference was easily explicable in terms of the correlation between velocity and brightness apparent in spectrograms such as Fig. 2.16. The central panel of Fig. 2.17 shows the individual profiles of granular and intergranular regions. The latter are redshifted with respect to the former and are less bright due to their lower temperature. The combination of shift and brightness difference reduces the contrast in the blue wing and enhances that in the red. On this basis, Beckers estimated the velocity difference between the two components to be 6 km s^{-1}.

In a subsequent paper, Beckers & Morrison (1970) analysed the filtergrams in more detail. They assumed that the line shape does not vary from point to point, and so were able to interpret the contrast $(I_b - I_r)/\frac{1}{2}(I_b + I_r)$ in terms of a simple line shift. They calibrated the relation between contrast and shift by tuning the filter through the line and measuring the contrast as a function of displacement from line centre. The line shift $\Delta\lambda$ then yielded a velocity $v = c\Delta\lambda/\lambda$, where c is the speed of light. The quality of the observations was insufficient for the study of individual granules, so they superposed some 1100 granules by aligning their brightest points. The composite images were then used to derive mean contrast and velocity profiles across the 'granule'. Moreover, by observing at different heliocentric angles (33°, 45°, 53°) they were able to estimate both the vertical and horizontal velocities.

Their composite granules showed an upflow at the centre and a downflow towards the boundary, a horizontal outflow from the centre, and no swirl (rotation about the vertical axis). This picture accords very well with our notions of a cellular convective flow. On the other hand, the numerical values quoted by Beckers & Morrison are subject to doubt for the following reasons:

(1) they assume that the line shape does not vary from point to point on the surface whereas there is, in fact, a systematic difference between granules and intergranular lanes with the result that the calibration is in error;

(2) the true centre of the granule is not known exactly for the analysis of the velocity measurements away from the centre of the disk. The

Fig. 2.17. Effect of granule and intergranule velocities on the granule contrast observed in a spectral line (schematic). The central panel shows the individual profiles, the right-hand panel the mean profile calculated on the assumption that the granule area is three times that of the dark lanes. It is evident that the contrast in the blue is much less than in the red. Moreover, the mean profile shows a marked asymmetry, illustrated by the curved bisector (Dravins, Lindegren & Nordlund, 1981).

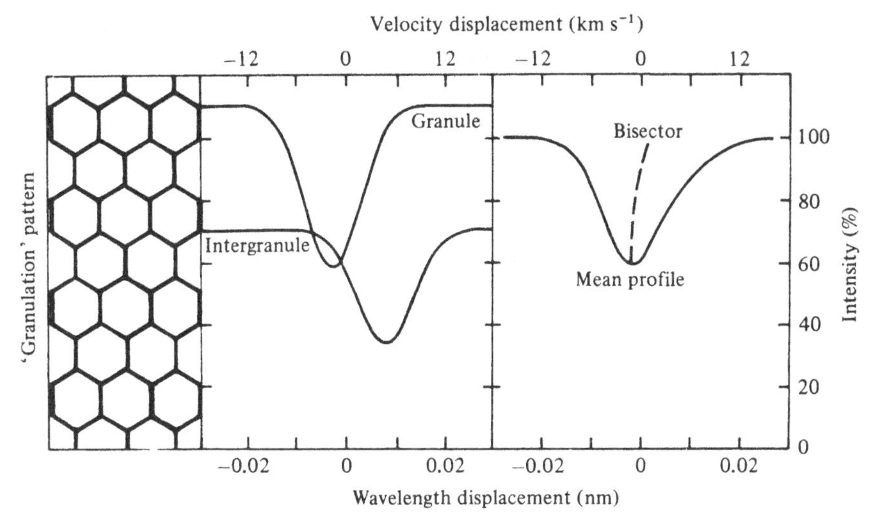

resolution of the line-of-sight velocity into horizontal and vertical components is affected by this choice;

(3) the seeing correction could only be estimated, with effects as in (2);

(4) most importantly, the superposition was performed without regard to the size and shape of the individual granules. Moving away from the centre of the composite granule, where the authors report a mean brightness excess of 5% and a mean upflow of 380 m s^{-1}, the properties of large granules become more and more mixed with those of the intergranular region surrounding small granules. Beckers & Morrison found the maximum outflow speed, 250 m s^{-1}, to occur at a distance of 480 km from the centre of the composite granule, where the cancellation effects are certainly significant and cause the outflow speeds to be underestimated.

It was not until 1976 that the first genuine measurement of individual granule velocities was made (Bray, Loughhead & Tappere, 1976). Using the 0.0125 nm filter installed on the 30-cm refractor of the CSIRO Solar Observatory, the Australian observers obtained a sequence of photographs at 0.005 nm intervals across the Fe I λ656.92 nm line within a 39 s period of good seeing. They then measured for ten typical granules the brightness ratio profile $R(\Delta\lambda) = I_i(\Delta\lambda)/I_g(\Delta\lambda)$ - referred to, rather confusingly, as a 'contrast' profile in the original paper - where I_g is the intensity at wavelength $\Delta\lambda$ from line centre in the middle of the granule, and I_i the corresponding intensity of the surrounding intergranular lane. By assuming that $R(\Delta\lambda)$ is the ratio of two Gaussian line profiles of different strengths and widths, with a relative Doppler shift, they were able to derive both the line parameters and the shift, making reasonable assumptions about the correction for atmospheric seeing. In this manner, they derived a velocity difference between the granule and the intergranular material of 1.8 km s^{-1} on the assumption that the velocities were substantially constant over the region of line formation. This measurement refers to the velocity difference between the centre of a granule and the (straight) intergranular lane and not the dark patches where several lanes meet; here the size of the downdrafts might be expected to be greater.

Keil (1980), on the other hand, sought to obtain information regarding the height variation of granular velocities from a series of spectrographic observations obtained at the Sacramento Peak Observatory. He measured the shifts of lines of increasing strength at points of brightness maxima and minima along the slit. Since the slit did not, in general, intersect the centres of granules, the true maximum and minimum granular velocities could be estimated only by means of a statistical correction. This drawback has to be weighed against the

Fig. 2.18. Granulation near west limb, photographed 0.01 nm redwards of the centre of the Fe I λ656.92 nm line with a 0.0125 nm filter on the 30-cm refractor of the CSIRO Solar Observatory, Culgoora (Loughhead & Bray, 1975). At this wavelength the granule contrast considerably exceeds the continuum value. Individual granules remain easily visible to within 5″ of arc of the limb; traces of inhomogeneous structure can still be discerned in the last 2″.

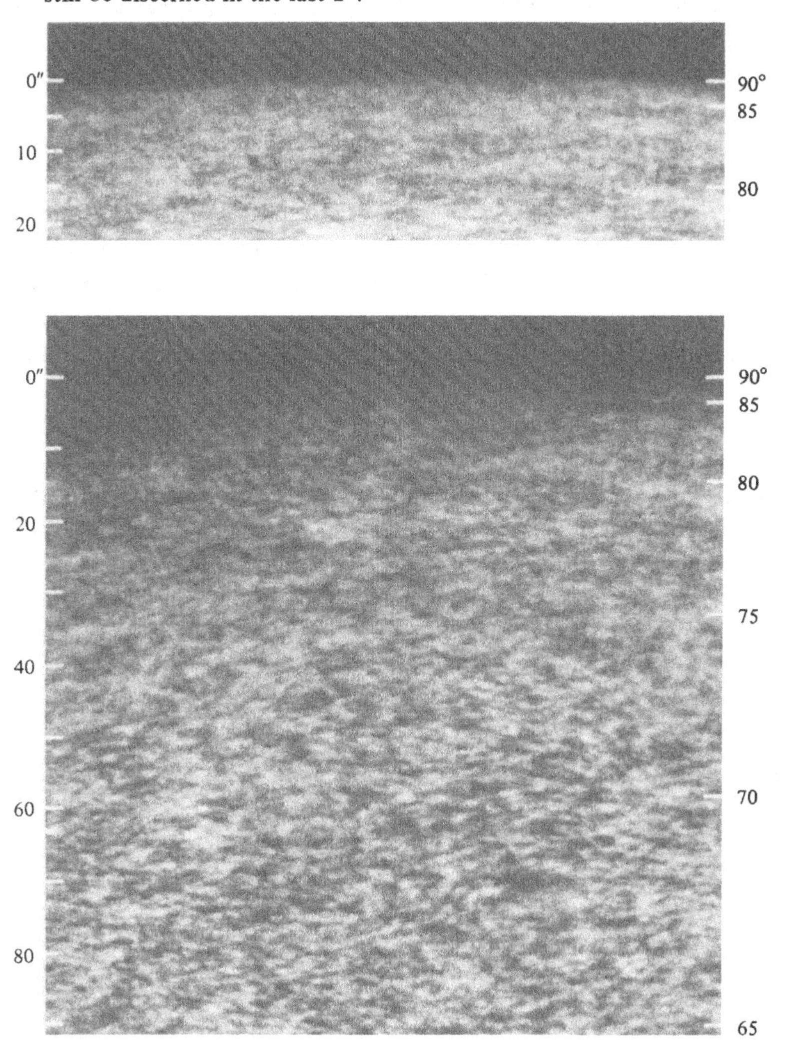

great advantage offered by such a time sequence which allows the small-scale oscillatory (non-granular) components of the photospheric velocity field to be filtered out (see Section 2.5.4). The mean amplitude of the line shift as a function of height (see Section 5.2.2) is shown in Fig. 5.2(*b*). The absolute values are subject to uncertainty due to the *ad hoc* correction procedures, but a sharp decline with height in the difference between the granular and intergranular velocities is clearly indicated.

Doppler shifts render the interpretation of filtergrams taken at a single wavelength in a wing of a spectral line ambiguous. Fig. 2.17 shows that differences of both intensity and velocity contribute to the brightness contrast. Both, of course, are manifestations of the processes underlying the granulation pattern. It is of great interest that Loughhead & Bray (1975) were able to detect features to within 5″ of the limb at the wavelength of maximum contrast in the filtergrams shown in Fig. 2.18 taken in the Fe I λ656.92 nm line. Even within the last 2″ from the limb they could detect inhomogeneities; these are considerably higher in the atmosphere than those revealed by the continuum observations discussed in Section 2.4.1.

The Australian observers (Bray, Loughhead & Tappere, 1974; Loughhead & Bray, 1975) also noted that the structure seen in filtergrams taken at the centre of the Fe I λ656.92 nm line differed from that visible in other parts of the line. Instead of the usual granulation pattern, a coarser, less sharply defined, bright and dark mottling was apparent. A pattern larger in scale than the granulation had earlier been noticed by Edmonds (1960, 1962b) on white-light Stratoscope photographs taken near the limb. This change from the normal granular structure deserves more attention than it seems yet to have attracted (see Sections 2.5.5 and 5.3).

2.5 Statistical properties of the solar granulation

2.5.1 Introduction

The photometric properties of the solar granulation discussed so far have been concerned almost exclusively with the maximum brightness or maximum velocity of individual granules. Apart from the description of Beckers & Morrison's work in the previous section, little has been said about the brightness and velocity distributions within the granules themselves. Such measurements pose a formidable observational task since the accuracy of observations of single granules is seriously limited by the instrumental and atmospheric resolution. One way to reduce the instrumental noise is to observe many granules and to average the results. This would be straightforward if the granulation were a strictly uniform and stationary pattern. In reality, however, the granulation pattern is constantly evolving and individual granules vary significantly in size, shape, and brightness (see Sections 2.3.5-2.3.7). It is steady in only the statistical sense; each granule is a different realization of the convective processes that underlie the whole phenomenon. In such a case, an analysis of the Fourier transforms of the brightness or velocity distributions, which are *mathematically* exactly equivalent to the distributions in physical space, has a much deeper significance. It provides estimates of the properties of the granulation treated as a stochastic field and, at least in principle, a means of establishing the confidence limits for the estimates.

Let us take the two-dimensional spatial brightness field $I(x_1, x_2)$ as an example. The complex Fourier transform is defined as

$$\mathscr{F}(I) = \int_{-\infty}^{\infty} I(x_1, x_2) \exp[\mathrm{i}(k_1 x_1 + k_2 x_2)]\, \mathrm{d}x_1\, \mathrm{d}x_2, \tag{2.4}$$

where k_1 and k_2 are the spatial wavenumbers in the x_1 and x_2 directions respectively. This function can be equally well described by two real functions, the amplitude A and the phase Φ:

$$\mathscr{F}(I) = A(k_1, k_2)\, \mathrm{e}^{\mathrm{i}\Phi(k_1, k_2)}. \tag{2.5}$$

Knowledge of A and Φ allows the original spatial field to be reconstructed exactly; the representations in physical space and Fourier space are equivalent. But if only the amplitude function or its square, the two-dimensional power spectrum,

$$P_{II}(k_1, k_2) = |\mathscr{F}(I)|^2, \tag{2.6}$$

is known, the reconstruction is not possible. On the other hand, the power spectrum has the important property of describing the contribution of each spatial scale to the total fluctuation of the brightness. A simple mathematical transformation shows that the power spectrum is just the Fourier transform of the autocovariance function

$$C_{II}(x_1, x_2) = \int_{-\infty}^{\infty}\!\!\int I(\xi_1, \xi_2)\, I(\xi_1 + x_1, \xi_2 + x_2)\, \mathrm{d}\xi_1\, \mathrm{d}\xi_2 \tag{2.7}$$

(see, for example, Jenkins & Watts, 1969). This function is the unnormalized form of the autocorrelation function and can be rigorously estimated from a finite sample if the field is statistically stationary and homogeneous. Away from active regions, the granulation fulfils these conditions; it is also isotropic, so the power spectrum is rotationally symmetric. With a change to polar coordinates $k_\perp$, ϕ, we can write

$$P_{II}(k_1, k_2)\, \mathrm{d}k_1\, \mathrm{d}k_2 = P_{II}(k_\perp)\, k_\perp\, \mathrm{d}k_\perp\, \mathrm{d}\phi, \tag{2.8}$$

where $k_\perp^2 = k_1^2 + k_2^2$. This reduces the power spectrum to a one-dimensional function which can be estimated by a one-dimensional sampling of the spatial brightness pattern. A scan in a fixed direction, the x_1-direction say, provides an estimate of the one-dimensional spectrum $P_1(k_1)$, which is related to the two-dimensional by

$$P_1(k_1) = 4\int_0^\infty P_{II}(k_1, k_2)\, \mathrm{d}k_2. \tag{2.9}$$

This equation can be inverted to yield the two-dimensional, rotationally symmetric, power spectrum through the Abel transform (Uberoi & Kovasznay, 1953),

$$P_{II}(k_\perp) = -\frac{1}{2\pi}\int_{k_\perp}^\infty \frac{\mathrm{d}P_1}{\mathrm{d}k_1}\frac{\mathrm{d}k_1}{(k_1^2 - k_\perp^2)^{1/2}}. \tag{2.10}$$

Statistical analysis is thus the natural treatment for the evaluation of spectra, which are perforce one-dimensional cuts and cannot provide a complete picture of an individual granule.

Working in the Fourier domain has two additional benefits. The first is purely practical. In Section 2.2 we described the smearing of optical images by the point-spread function of the telescope and atmosphere. Mathematically, this smearing is a convolution, which in physical space is equivalent, by Parseval's theorem, to a simple product of the relevant transforms in Fourier space. The amplitude of the point-spread function is known as the modulation transfer function or MTF. The power spectrum of the true brightness distribution is thus obtained by simply dividing the observed power spectrum by the square of the MTF, so long as the MTF remains finite. The computational ease offered by working in Fourier space is very marked, especially when the point-spread function is not a simple analytic function (Ricort & Aime, 1979).

The second benefit of the Fourier transform approach is the link to theory that it provides. Taking the Fourier transform decomposes the brightness field into different horizontal modes. In a stably stratified atmosphere, these modes correspond to natural wave modes which are independent of one another if the atmosphere is isothermal (see Bray & Loughhead, 1974: Section 6.3). Although the solar photosphere is not strictly isothermal, it approximates that condition closely enough for the assumption of independent modes to be useful. In the convectively unstable regions of the atmosphere, some wave modes are replaced by convective modes corresponding to each horizontal wavenumber. In finite-amplitude convection these modes are not independent; however, several attempts have been made to simplify the physics of the convective process by isolating a single dominant mode (see Section 4.3.2).

It should be stressed that the choice of horizontal wavenumber alone does not suffice to specify uniquely the mode of the wave or of the convection. We require also the spatial dependence of the fluctuation on height or, equivalently, the temporal dependence. This behaviour can again be studied by means of a Fourier decomposition - in this case into various temporal frequencies ω - if a time series of observations is available. The diagnostic importance of knowing the k and ω dependence of solar phenomena can be readily appreciated; it allows us to recognize and distinguish the simultaneous presence of different physical processes. It has been a particularly useful tool in the investigation of the nature of the velocity field in the solar atmosphere (see Section 2.5.4).

2.5.2 *Measurement of the granular brightness distribution*

The various modes of the brightness and velocity fluctuations in the solar photosphere can be seen clearly in the analysis of Deubner (1974b). Utilizing a long time series of spectra with exceptional spatial resolution obtained at Sacramento Peak Observatory, he measured the brightness fluctuations and Doppler shifts of the very weak, low photospheric line C I λ538.03 nm and of the strong, upper photospheric line Fe I λ538.34 nm. The results are displayed in so-called diagnostic diagrams which show the fluctuation power as a function of horizontal wavenumber $k_\perp$ and temporal frequency ω. The diagrams for the brightness variations are shown in Fig. 2.19(*b*) and (*d*). They reveal that the brightness power throughout the photosphere is distributed almost exclusively along the wavenumber axis, i.e. at almost vanishing temporal frequencies. This simply reflects the fact that the brightness pattern is long-lived and aperiodic; it contains only a convective contribution. It is also noteworthy that the power is concentrated not only at granular scales ($1.5 < k_\perp < 6\,\mathrm{Mm}^{-1}$ or, in terms of spatial wavelengths $\Lambda_\perp$, $1''.5 < \Lambda_\perp < 6''$) but also at larger scales ($k_\perp < 1\,\mathrm{Mm}^{-1}$ or $\Lambda_\perp > 9''$), especially in the upper photosphere. This large-scale brightness pattern can be identified with the photospheric network (see Section 2.6 and Fig. 2.14(*b*)), which is cospatial with the supergranulation (see Section 2.7). This network has a cell size of some 48″ but is very irregular; these irregularities show up in the power spectrum at smaller scales. To isolate the granulation we need only to filter out the spatial brightness fluctuations that are larger than 8″ (i.e. wavenumbers smaller than $\sim 1\,\mathrm{Mm}^{-1}$).

The many recent measurements of granular brightness fluctuations are summarized in Table 2.9. Two features are worthy of remark. Firstly, exceptional spectrograms can show root mean square (rms) values similar to those of the best white-light photographs, although the much longer exposure times required generally result in greater atmospheric smearing and reduced values. Secondly,

Fig. 2.19. Diagnostic ($k_\perp$, ω) diagrams of velocity and brightness fluctuations in two photospheric lines measured at disk centre. The carbon line is formed in the low photosphere, the neutral iron line in the upper photosphere. The power is indicated by shading, the limiting levels being 2.8, 25 and 69% (Deubner, 1974b).

(*a*)

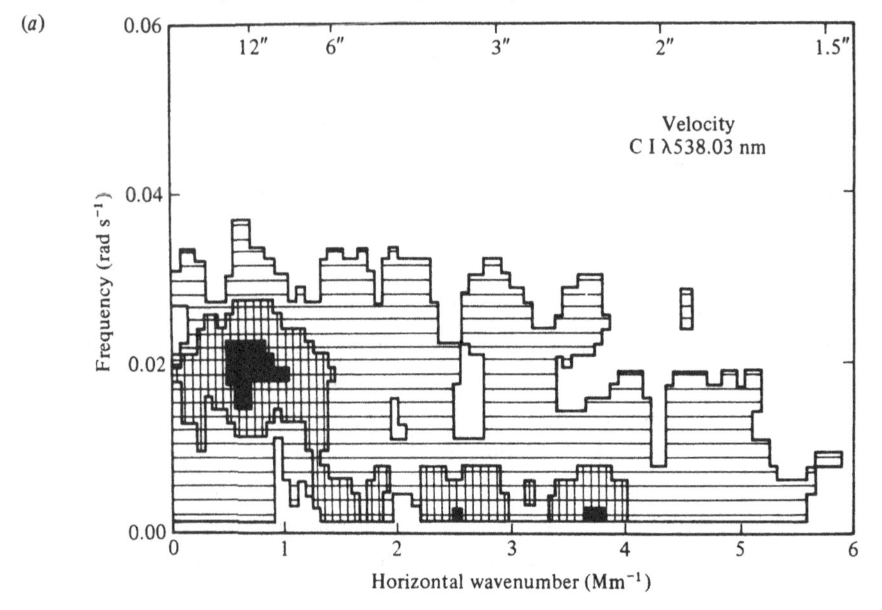

(*b*)

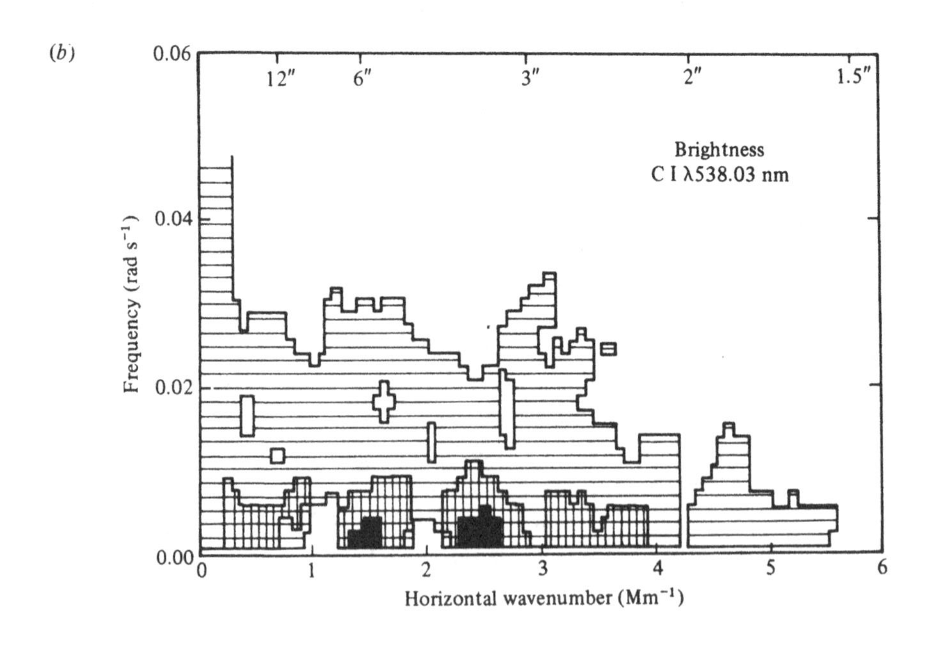

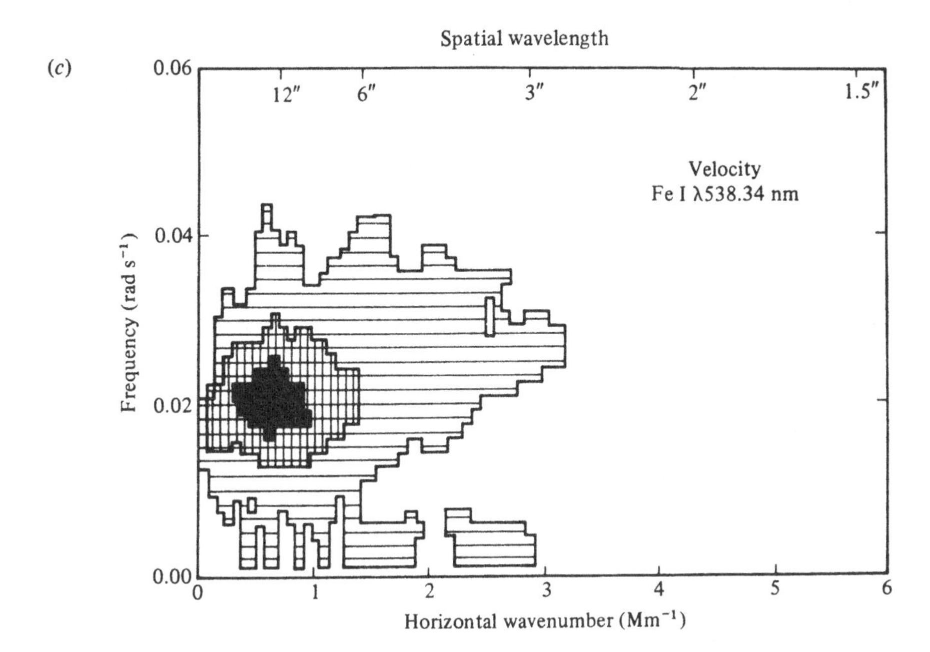
(c)
Spatial wavelength
12″
6″
3″
2″
1.5″
Velocity
Fe I λ538.34 nm
Frequency (rad s⁻¹)
0.06
0.04
0.02
0.00
0
1
2
3
4
5
6
Horizontal wavenumber (Mm⁻¹)

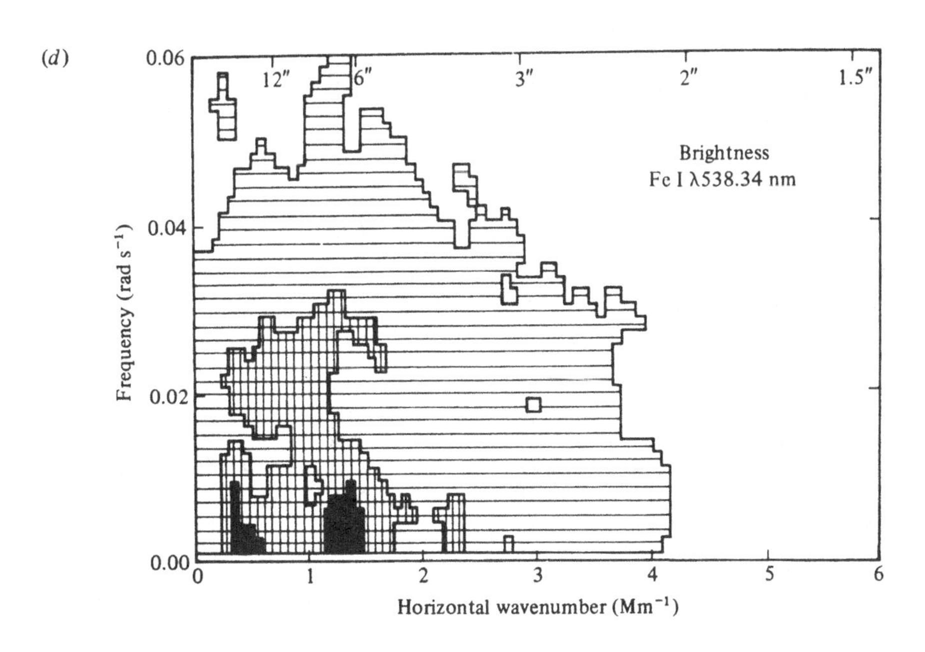
(d)
12″
6″
3″
2″
1.5″
Brightness
Fe I λ538.34 nm
Frequency (rad s⁻¹)
0.06
0.04
0.02
0.00
0
1
2
3
4
5
6
Horizontal wavenumber (Mm⁻¹)

the balloon-borne telescopes have yielded the largest uncorrected rms values. Nevertheless, they are not ideal for the measurement of the granular brightness owing to

(1) the comparatively small apertures of the telescopes;
(2) difficulties of photometric calibration;

Table 2.9. *Measurements of rms brightness fluctuation*

Reference	Type of observation D: direct photograph S: spectrogram	Wave-length (nm)	$\langle I'/\bar{I} \rangle$ (%) Measured	Corrected
Balloon observations				
Blackwell *et al.* (1959) (corrected Gaustad & Schwarzschild, 1960)	D	530	4.6	7.3
Schwarzschild (1959)			4.6	7.3
Bahng & Schwarzschild (1961)			5.0	7.2
Edmonds (1962b)	D	548	5.5–8.9	9.2–15
Leighton (1963)			5.2–7.2	9.3–13
Namba & Diemel (1969)			5.6	
Edmonds & Hinkle (1977)			7.4–8.5	13.6–14.5
Krat *et al.* (1972)	D	465	8.8	12–14
Wittmann & Mehltretter (1977)	D	556	5.7	8.7
Schmidt *et al.* (1979)			4.05	8.6
Wittmann (1981)			6.2	8.9
Ground observations				
Evans & Michard (1962a)	S	517	3.05	
Mehltretter (1971a)	S	630	2.6	
Mehltretter (1971b)	D	552	5.9	
Lévy (1971)	D	530	7.0	17.5
Canfield & Mehltretter (1973)	S	517	5.85	
Deubner & Mattig (1975) (corrected Schmidt *et al.*, 1979)	D	607	5.0	11.5–12.0
Keil (1977)	D	552	5.7	
Keil & Canfield (1978)	S	518	5.0	
Schmidt *et al.* (1979)	D	422	4.9	10.4
Ricort & Aime (1979)	D	505	1.3–7.25	17.2
Keil (1980)	S	520	4–6	
Schmidt *et al.* (1981)	D	550	4.2–6.35	
Ricort *et al.* (1981)	D	530		17.9

(3) residual seeing even at balloon altitudes;
(4) degradation of optical performance during the flight.

A small aperture means that the instrumental corrections are large, while calibration difficulties make them uncertain. The balloon results are included in Table 2.9 but, with the possible exception of those from the Soviet SSO (Krat *et al.*, 1972), the values should be given less weight than carefully calibrated and corrected data from ground-based observations; we shall discuss exclusively the latter in the following.

The first such analysis was undertaken by Lévy (1971) using continuum photographs obtained at Pic-du-Midi at the time of a solar eclipse. When the Moon partially covers the solar disk, seeing and diffraction cause light to spill over the sharp edge of the lunar limb. The resulting intensity distribution as a function of distance from the limb yields immediately the one-dimensional form of the point-spread function, the line-spread function (LSF). Lévy converted one-dimensional scans of the granular brightness distribution to a two-dimensional form by means of the Abel transform, and applied a seeing correction in the form of an analytic approximation to the LSF. This increased the rms relative brightness fluctuation $\langle I'/\bar{I} \rangle$ in the wavenumber range $0 < k_\perp < 6.0\,\mathrm{Mm}^{-1}$ from the observed value of 7%, itself a very high value, to 17.5%.

The conversion of one-dimensional power spectra to a two-dimensional form requires the statistical properties of the granulation (the Fourier amplitudes but not phases) to be rotationally invariant. The validity of this assumption has been demonstrated by several authors (Deubner & Mattig, 1975; Karpinsky & Mekhanikov, 1977; Edmonds & Hinkle, 1977) and is not a source of significant error. However, Lévy's choice of analytical approximation to the LSF was later criticized by Deubner & Mattig (1975). These authors again used material from a partial eclipse, observed at Izaña at a wavelength of λ607 nm. A print of one of the frames is reproduced in Fig. 2.4(*b*), showing part of the Moon occulting the solar disk. The two-dimensional brightness spectrum was corrected using both the numerical values of the LSF derived from lunar limb scans and various analytic representations of it. The corrections were found to be very sensitive to the fit in the wings of the LSF. The two-component Gaussian model used by Lévy led to large errors.

Deubner & Mattig derived a corrected rms value of 12.8% from their measured value of 5% for $0 < k_\perp < 12\,\mathrm{Mm}^{-1}$. This figure was subsequently corrected for photographic noise by Schmidt, Deubner, Mattig & Mehltretter (1979) and revised to 11.5-12%.

A different method of deriving the seeing corrections has been developed by Ricort & Aime (1979) (see Section 2.2.3). It depends upon the assumption of a model for the turbulent phase fluctuations in the wave front. Using the model of Korff, Dryden & Miller (1972), they were able to determine Fried's parameter - the effective diameter of the telescope in the presence of atmospheric degradation - for each of a series of one-dimensional scans of the granulation. The individual estimates of $\langle I'/\bar{I}\rangle$ ranged from 1.3% to 7.25%; the mean corrected value was 17.2%, integrated over the range $0 < k_{\perp} < 22\ \mathrm{Mm}^{-1}$. These observations were made at Sacramento Peak Observatory at λ505 nm. Very similar results were obtained by the same authors, together with Deubner & Mattig, from observations of lower quality obtained at the Capri Observatory during another partial eclipse (Ricort, Aime, Deubner & Mattig, 1981). The MTF - in this case the Fourier amplitude function of the LSF - was determined, as before, from the lunar limb and found to be in accord with the atmospheric turbulence model. When integrated as far as $k_{\perp} = 17\ \mathrm{Mm}^{-1}$, $\langle I'/\bar{I}\rangle$ was equal to 18% at λ530 nm.

Fig. 2.20 compares the two-dimensional spectrum of Ricort & Aime with that of Deubner & Mattig. The French authors were able to exploit the large (76-cm) aperture of the tower telescope at Sacramento Peak Observatory in

Fig. 2.20. Two-dimensional power spectra of granule brightness fluctuations (Ricort & Aime, 1979). The measurements of Ricort & Aime and of Deubner & Mattig (1975) are corrected for instrumental and atmospheric blurring, but those from the SSSO are not (Karpinsky & Mekhanikov, 1977). The curves illustrate the substantial loss of resolution evident even in a telescope of 50-cm aperture (SSSO).

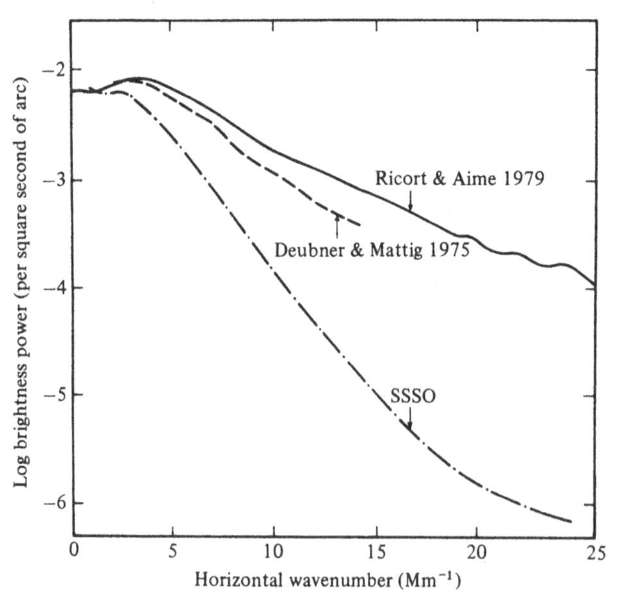

order to extend the brightness spectrum as far as spatial wavenumbers of 26 Mm^{-1} (0\".33). The more modest 45-cm aperture of the Izaña telescope results in a cutoff at 15 Mm^{-1} (0\".6). Nevertheless, the general agreement between the curves in their respective regions of reliability is good. In particular, the roughly exponential decay of the power at high wavenumbers found by Deubner & Mattig is confirmed and found to extend beyond 20 Mm^{-1}, i.e. below spatial scales of 0\".4. Some impression of the magnitude of the corrections implied by these results may be gained by comparing the 'true' spectra with the *uncorrected* spectrum measured from the SSSO flight with a telescope of 50-cm aperture (Karpinsky & Mekhanikov, 1977). At 20 Mm^{-1} (0\".4) the discrepancy exceeds two orders of magnitude. The difficulty attending accurate photometry of the intergranular lanes cannot be more forcefully demonstrated (see Section 2.3.4).

To summarize, the best estimate of the total rms relative brightness fluctuation at λ500 nm appears to be 15-17%.

The values of $\langle I'/\bar{I}\rangle$ that measurements with different spatial cutoffs would yield can be easily calculated from the power spectrum, since the mean square fluctuation is simply the area under the curve between the appropriate spatial limits. If we accept the corrected curve of Ricort & Aime as a basis for comparison, we find from Table 2.10 that $\langle I'/\bar{I}\rangle$ increases from about 8% to 15% as the spatial cutoff decreases from 1\".8 to 0\".7. 25% of the power is contributed by scales between 1\".8 and 5\".4, i.e. by scales greater than the characteristic cell size of the granulation (Section 2.3.3). 50% of the power is reached at scales of 5.9-6.1 Mm^{-1}, corresponding to a wavelength of about 1\".4. This is somewhat less than the mean cell size of 1\".9 obtained by direct measurement (cf. Section 2.3.3), but this is to be expected since the Fourier analysis does not distinguish between granules and granular fragments. When granular fragments are included in a direct measurement (Namba & Diemel, 1969), the result is a mean cell size of 1\".3-1\".4, in reassuring agreement.

Statements about the shape of the power spectrum raise the question of the statistical significance of the estimates. This has been discussed by, amongst

Table 2.10. *Apparent rms brightness fluctuation (λ505 nm) as a function of spatial cutoff*

Spatial wavenumber range, $k_\perp$ (Mm^{-1})	Spatial wavelength range, $\Lambda_\perp$ (seconds of arc)	$\langle I'/\bar{I}\rangle$ (%)
$1.6 < k_\perp < 3.3$	$5.3 > \Lambda_\perp > 2.7$	5.0
$1.6 < k_\perp < 4.9$	$5.3 > \Lambda_\perp > 1.8$	8.4
$1.6 < k_\perp < 6.5$	$5.3 > \Lambda_\perp > 1.33$	10.9
$1.6 < k_\perp < 13.0$	$5.3 > \Lambda_\perp > 0.67$	15.2

others, Deubner (1974a) and Keil (1977). There seems to be general agreement that the brightness power spectrum shows two main peaks at 3.4 and 1.7 Mm^{-1} (2.″5 and 5″), which are present in Fig. 2.19(*b*). The latter peak lies significantly outside the range of single granule scales. Variations in the granulation *pattern* both between frames and within single frames obtained by Stratoscope I were noted by Edmonds & Hinkle (1977). Both this phenomenon and the 5″ peak in the power spectrum can probably be attributed to topological clumping of granules, i.e. the tendency to form families noted in Section 2.3.7. The multiple peaks in the granular power spectra at higher wavenumbers reported by Mattig & Nesis (1974) and Wittmann (1981) are not so well substantiated. They are certainly present in individual samples, but it is doubtful whether they are an essential property of the granulation as a whole (Deubner, 1974a).

2.5.3 *Variation of the granular brightness fluctuation with heliocentric angle and wavelength*

There are few systematic investigations of the variation of granular brightness fluctuations with heliocentric angle θ and with wavelength. Such variations are small and can easily be falsified by errors in the correction procedure. This was the fate of the first determination of the centre-to-limb variation of the granular brightness fluctuation from the Stratoscope material by Edmonds (1962b); it showed a narrow maximum at $\theta = 50°$ ($\mu = 0.64$). A similar difficulty arose in the analyses by Schmidt *et al.* (1979) who utilized data from both Spektrostratoskop and a partial eclipse (cf. Section 2.5.2); their published values should be divided by a factor of $\mu^{1/2}$ (Wiesmeier & Durrant, 1981). Table 2.11 lists the corrected values of the rms relative brightness fluctuation $\langle I'/\bar{I}\rangle$, integrated over the wavenumber range $0 < k_\perp < 10\,Mm^{-1}$, as a function of the

Table 2.11. *Variation with heliocentric angle of brightness fluctuation and mean wavenumber*

μ	$\dfrac{\langle I'(\mu)/\bar{I}(\mu)\rangle}{\langle I'(1)/\bar{I}(1)\rangle}$	$\bar{k}_\perp$ (Mm^{-1})
1.0	1.0	5.9
0.86	1.04	5.8
0.64	0.93	5.7
0.54	1.00	5.2
0.30	0.71	4.6
0.20	0.56	4.8

heliocentric angle; they reveal a broad maximum at the centre of the disk. A very similar dependence on heliocentric angle was found by Pravdjuk *et al.* (1974) using SSSO data and by Keil (1977).

The monotonic decrease in $\langle I'/\bar{I}\rangle$ with increasing heliocentric angle is now well established. This decrease towards the limb is accompanied by an increase in the horizontal scale of the pattern. Schmidt *et al.* (1979) found that the mean wavenumber decreases from 5.9 Mm^{-1} (1″.45) at the centre of the disk to 4.8 Mm^{-1} (1″.8) at $\mu = 0.2$ (see Table 2.11).

To interpret the measurements let us make the simple assumption that the observed radiation arises from the level of the solar atmosphere where an optical depth of μ is reached (the Eddington-Barbier approximation, see Section 4.4.1). Thus

$$I_\lambda = B_\lambda(\tau_\lambda = \mu). \tag{2.11}$$

By appealing to Wien's approximate form of the black-body radiation law

$$B = (2hc^2/\lambda^5)\, \mathrm{e}^{-hc/(k\lambda T)}, \tag{2.12}$$

we obtain the relative brightness fluctuation (Schröter, 1957)

$$I'/\bar{I} = B'/\bar{B} = -(hc/k\lambda)\, T'(\tau_\lambda = \mu)/\bar{T}^2. \tag{2.13}$$

Over a sufficiently small wavelength range the brightness fluctuations vary, in this approximation, only with the atmospheric temperature fluctuations at optical depth μ. Fig. 2.21 shows examples of the rms relative brightness fluctuation $\langle I'/\bar{I}\rangle$, expressed as a fraction of the value in the continuum at the centre of the disk, as a function of the height where $\tau = \mu$. The data all stem from Spektro-stratoskop measurements, the continuum values from Schmidt *et al.* (1979) and the values for the central intensity fluctuations of various Fraunhofer lines from Maluck (1980); there are no systematic differences between these two sources. At the centre of the disk, $\langle I'/\bar{I}\rangle$ is almost constant up to a height of 200 km. The curves corresponding to the heliocentric angles $\mu = 0.6$ and 0.2 are, however, progressively depressed below that of the centre of the disk. Since the temperature at a given nominal height is an isotropic scalar quantity, it cannot depend on the inclination of the line of sight. The discrepancy must therefore arise from geometrical effects introduced by oblique lines of sight cutting through more than one structure near the limb. This observation reinforces the conclusion of Section 2.4.1 that the interpretation of limb observations requires the theory of radiative transfer in a non-uniform medium (see Sections 4.4.3 and 5.2.2).

Let us now consider the variation of rms brightness with wavelength at the centre of the disk. Eqn. 2.13 predicts that $\langle I'/\bar{I}\rangle$ should be inversely proportional to the wavelength. This is borne out by the rather limited observational results available. The Soviet SSO obtained spectrograms simultaneously in various spectral regions from which $\langle I'/\bar{I}\rangle$ was derived at several wavelengths by Karpinsky & Pravdjuk (1972). Unfortunately, the spatial resolution achieved does not approach that in the white-light pictures; the uncorrected $\langle I'/\bar{I}\rangle$ at λ580.0 nm is only 3.3%. These measurements have been extended into the infrared by Albregtsen & Hansen (1977), but the observations, made at Oslo, were obtained with a photometer aperture of diameter 1–2″; thus $\langle I'/\bar{I}\rangle$ was only 1.7% at λ578.0 nm. Both sets of measurements, shown in Fig. 2.22, thus reflect mainly the contribution of the large granules. The departures from the Wien curve, the line of slope minus one, cannot be considered significant.

We can now use Eqn. 2.13 to compare the two most reliable determinations of the rms relative brightness fluctuations. Ricort & Aime found a value of

Fig. 2.21. Variation of the rms brightness fluctuation with residual line intensity and height. Filled symbols show values for the cores of various Fraunhofer lines at different heliocentric angles, open triangles values for the continuum. Heights are based on the Eddington–Barbier relation (see text) (Durrant, Kneer & Maluck, 1981).

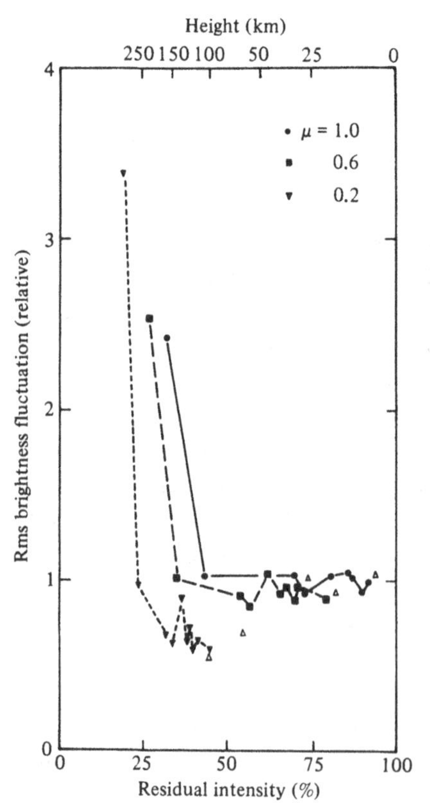

15.2% at λ505.0 nm for $1.6 < k_\perp < 12\,\mathrm{Mm}^{-1}$ (Table 2.10). Converting this to λ607.0 nm, we obtain a value of 12.6%, which is in good agreement with the value of 11.5-13.0% obtained by Deubner & Mattig.

If the granulation brightness distribution were flat-topped, an rms value of 12% at λ550 nm would correspond to a contrast of 24%, while if it were sinusoidal in two dimensions then the contrast would be 48%. The contrast directly measured by Bray & Loughhead with a comparable resolution at λ550.0 nm is 24% and by Karpinsky at λ465 nm is 36%, equivalent to 30% at λ550 nm (Section 2.3.4). This implies a flat-topped brightness distribution in agreement with the appearance of good granulation photographs (cf. Section 2.1).

2.5.4 Velocity distributions in the solar photosphere

The statistical properties of the velocity field associated with the granulation are not known reliably for four reasons, three of which we have met before. Firstly, the observations suffer from the statistical uncertainty inevitably associated with the limited samples provided by spectrograms. Secondly, the method for correcting the measured line shifts for seeing effects is only valid for shifts small in comparison with the Doppler width of the lines (Mehltretter, 1973), and this criterion is barely satisfied by granular velocities.

Fig. 2.22. Wavelength variation of the rms brightness fluctuation of the granulation. The measurements of Karpinsky & Pravdjuk (1972) (circles) and Albregtsen & Hansen (1977) (squares) have been scaled to bring them into coincidence in the visible region. The line shows the relationship calculated from the Wien approximation to Planck's law assuming a temperature fluctuation independent of height.

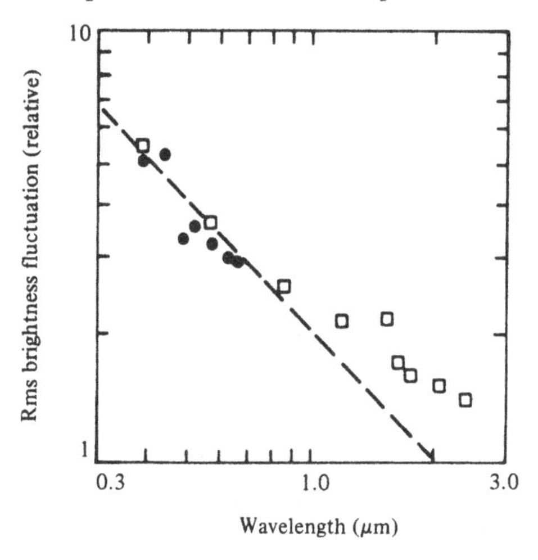

Thirdly, the velocity field is a vector field, so that its vertical and horizontal components must be distinguished. Finally, the granulation velocities are masked by other velocity fields of comparable magnitude.

The recognition of the last problem stems from the powerful attack on the question of granule velocities initiated in 1960 when Leighton, working at the Mt Wilson Observatory, developed an ingenious new technique for 'photographing' line-of-sight velocities with the aid of an ordinary spectroheliograph over the whole, or any selected portion, of the solar disk.

Using the new method Leighton and his collaborators immediately established that the velocity field in the upper layers of the photosphere actually consists of two physically distinct regimes: one is a small-scale field of upward and downward vertical motions, and the other is a large-scale pattern of horizontal motions constituting the phenomenon of the supergranulation described in Section 2.7.

The spectroheliographic method is especially adapted to the task of studying temporal changes in velocity fields and was used by Leighton, Noyes & Simon (1962) to make the first systematic study of the time variation of the small-scale photospheric velocity field. This led them to the striking discovery that the small-scale field is markedly oscillatory in character, the pattern of velocities varying almost sinusoidally in time with a period very close to 5 min. Confirmation of the discovery was quickly afforded by observations by Evans & Michard (1962b) and by Howard (1962), using in each case an entirely different observational technique.

In recent years, thanks to the observational work of Deubner and the theoretical work of Ulrich (1970) and of Ando & Osaki (1975), the nature of these oscillations has become clear. They are acoustic waves trapped in the outer envelope of the Sun. The upward and downward propagating waves interfere and set up standing (global) oscillations. The discrete character of the modes was first demonstrated from a very long series of observations by Deubner (1975) and was confirmed by Rhodes, Ulrich & Simon (1977). For shorter strings of data, the modes are not resolved but appear as a broad band of power in the ($k_\perp$-ω) diagram centred on a period of 5 min and extending over the range $0.5 < k_\perp < 1\ \mathrm{Mm}^{-1}$ ($8'' < \Lambda_\perp < 16''$); they are clearly recognizable in the diagnostic diagrams of Fig. 2.19.

The diagram for the very weak, low photospheric carbon line shows, as well as the oscillatory power, considerable power at lower frequencies and higher wavenumbers; this is the level where the brightness fluctuations of the granulation occur. In the core of the strong iron line, on the other hand, there appears to be almost no velocity power that could be called granular.

Looking again at the diagram for the velocities in the low photosphere, we see that the oscillatory and convective motions do not separate cleanly in wavenumber. For the wavenumber range $1 < k_\perp < 2.5\,\mathrm{Mm}^{-1}$ ($3''.5 < \Lambda_\perp < 8''.5$) both types of motion are present with about equal power. It is thus impossible to separate the components on the basis of spatial characteristics alone - in particular, on a single spectrogram. Unfortunately, most authors are forced to work with single spectrograms since very high-quality time series are almost non-existent. Present results are not surprisingly very discordant.

To our knowledge there has been only one attempt to derive granular velocities from material that allowed a simultaneous determination of the line-spread function of the telescope and atmosphere. Durrant *et al.* (1979) used spectra taken during an eclipse with a radial slit crossing the Moon's limb and applied to the Doppler shift measurements the seeing correction appropriate to the brightness field. This is valid in the limit of small velocities (Mehltretter, 1973). The granulation was separated from the oscillations by excluding all velocity power with wavenumbers less than $2.3\,\mathrm{Mm}^{-1}$ (i.e. $\Lambda_\perp > 3''.8$). This is a conservative choice of cutoff: to be certain of excluding the oscillations, some granular power is sacrificed and the resulting values therefore represent lower bounds. There is a difference of a factor of four between the observed and corrected velocities, highlighting the uncertainty in working with material obtained under only average seeing.

Canfield and his coworkers were fortunate in obtaining high-quality material. As may be seen from Table 2.9, the exceptional spectrograms obtained at Sacramento Peak Observatory (Canfield & Mehltretter, 1973; Keil & Canfield, 1978; Keil, 1980) approach the quality of stratospheric white-light photographs. This minimizes the seeing corrections, but they are nevertheless still not negligible and remain unknown. In addition, Canfield and his collaborators used a different technique for separating oscillations from granulation. The oscillatory velocity field is known to increase in amplitude with height, so they assumed that a sharp decrease of velocity with height distinguishes granular motions from the oscillations. This suggests fitting the observed rms velocity fluctuations as a function of height x_3 by the two exponential curves

$$\langle v'_{3,\mathrm{osc}}\rangle = \langle v^0_{\mathrm{osc}}\rangle \exp(x_3/H_{\mathrm{osc}})$$
$$\langle v'_{3,\mathrm{gr}}\rangle = \langle v^0_{\mathrm{gr}}\rangle \exp(-x_3/H_{\mathrm{gr}}),$$

where the scale height for the oscillations, 1100 km, was determined from the observations of Canfield & Musman (1973). The best fit for the remaining parameters yielded $\langle v^0_{\mathrm{gr}}\rangle = 1.27\,\mathrm{km\,s^{-1}}$, $H_{\mathrm{gr}} = 150\,\mathrm{km}$ (Canfield, 1976) and $\langle v^0_{\mathrm{gr}}\rangle = 1.45\,\mathrm{km\,s^{-1}}$, $H_{\mathrm{gr}} = 140\,\mathrm{km}$ (Keil & Canfield, 1978).

These results differ markedly from those of Durrant *et al.* This can be seen in Fig. 2.23, which shows the derived variation of the granular and the oscillatory rms velocity fields as a function of height. The results of Durrant *et al.* imply that the magnitude of $\langle v'_{3,\mathrm{gr}}\rangle$ hardly changes with height, dropping from about

0.9 km s^{-1} at the base to some 0.7 km s^{-1} at 500 km (i.e. the level of the temperature minimum). Canfield's results indicate that the motions are rapidly damped from a value of about 1 km s^{-1} at the base to almost vanishing values at 500 km.

Keil (1980) sought to resolve this difference with the aid of a time series of very high-quality spectrograms obtained at Sacramento Peak Observatory. During the 30 min of observation the continuum $\langle I'/\bar{I}\rangle$ maintained the high average value of 5.2%. This enabled him to obtain the diagnostic diagrams for the Doppler shifts of lines of different strengths and thus to distinguish the true granular component from the oscillations. He found that the granular component was much weaker in lines formed in the middle photosphere (at a height of 200 km) than in lines formed in the low photosphere. It was absent in lines formed in the high photosphere, more than 500 km above $\tau = 1$.

On the basis of Keil's results it seems safe to conclude that the small-scale power in strong lines (at heights of 500 km) should not be attributed to the

Fig. 2.23. Height variation of the rms vertical granule and oscillatory velocities. The exponential and linear representations are due to Canfield (1976) and Durrant *et al.* (1979) respectively. Dotted segments are extrapolations.

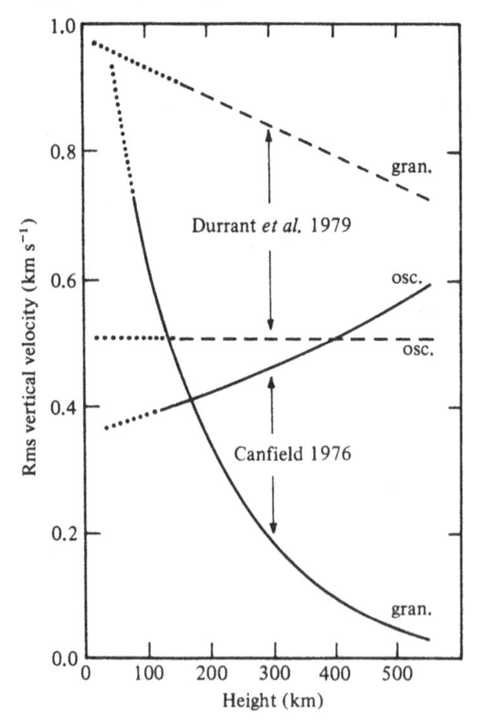

granulation but to oscillations (see Fig. 2.19). In medium-strength lines, the picture is not so clear: earlier observations by Frazier (1968) suggest more long-period power than was found by Keil. Some short-period velocity fluctuations certainly contribute to the power measured in these lines by Durrant *et al.*, but the true gradient of the rms vertical velocity up to heights of 500 km is still uncertain.

It is important to find the reason for these discrepancies. The decline of the convective velocity field in the stably stratified layers of the atmosphere is very significant for theories of convective overshoot, particularly for the evaluation of semi-empirical models of the granulation (see Section 5.2.3).

The centre-to-limb variation of the rms velocity is even more uncertain. The correction for the seeing and instrumental degradation is very problematical. This difficulty is all the more serious as the corrections change syste-

Fig. 2.24. Variation of rms total velocity (granule + oscillatory) at disk centre ($\mu = 1.0$) and at $\mu = 0.8$ with residual line intensity and height. The dashed line represents the oscillatory component in the upper photosphere. Note that the velocity in the low photosphere is greater at $\mu = 0.8$ than at $\mu = 1.0$; this could indicate that the horizontal granule velocity is larger than the vertical (after Keil & Canfield, 1978).

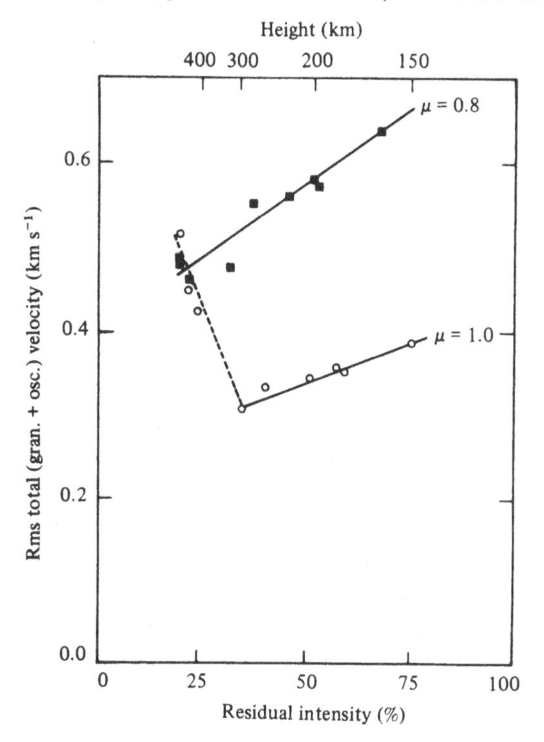

matically from centre to limb, so that an incorrect restoration will produce not just scaling errors but a distorted variation with heliocentric angle.

In the absence of reliable results, we shall merely quote uncorrected data from published sources. In Fig. 2.24 the rms total velocities for various lines at two heliocentric angles measured by Keil & Canfield (1978) are compared. An effective height for each line is indicated by the centroid of the velocity weighting function (cf. Section 4.4.1) and should be interpreted as only a rough guide to the atmospheric layer to which the velocity measurement refers.

The measured rms velocities at a heliocentric angle of 37° are about 1.6 times greater than those at disk centre at the same effective height, suggesting that the horizontal component is roughly twice the vertical component. A more detailed analysis, such as that presented by Keil & Canfield, requires a model for the other velocity fields present - the oscillations and the supergranulation - but leads to essentially the same result. This result is supported by the analysis of Spektrostratoskop observations by Mattig, Mehltretter & Nesis (1981).

2.5.5 *Correlations*

The techniques described in the previous two sections can oe used not only to determine the variation of the brightness and velocity fluctuations with height but also to study their correlation as a function of height. This provides insight into such important theoretical questions as the height to which the convective motions extend and the amount of heat they transport.

Richardson & Schwarzschild (1950) used a Mt Wilson spectrogram showing fine definition to compare Doppler displacements along the length of the slit with the brightness fluctuations at the corresponding points in the continuum. They found that the correlation between the velocity and brightness was rather weak, only the narrow regions of high upward velocity appearing to be systematically brighter than average.

Richardson & Schwarzschild's data were later carefully re-examined by Stuart & Rush (1954). These authors formed the opinion that there was actually a good correlation between the brightness fluctuations produced by the photospheric granulation and small-scale variations in velocity, but that this correlation was masked by large-scale fluctuations in velocity which did not appear to be closely correlated with brightness. To test this hypothesis they first calculated the coefficient of correlation r between the deviations of velocity and brightness from the overall mean values given by Richardson & Schwarzschild. The resulting value, $r = -0.30$, indicated only a weak correlation, the negative sign implying that bright areas tend to move upwards and dark areas downwards. Stuart & Rush then used the technique of 'moving-averages' to remove all variations in velocity and brightness on a scale exceeding about 5000 km. Following this they re-computed

r in terms of the deviations from the moving averages, obtaining the value $r = -0.68$. The new figure indicated a strong correlation between velocity and brightness on a scale corresponding to the scale of the granulation pattern.

The large-scale velocity field is sensibly constant across an individual granule, so that the relative Doppler shift between the centre of the granule and its surrounding intergranular material is unaffected – hence the 100% correlation between the *sense* of motion and brightness found by Bray *et al.* (1976).

This conclusion is reinforced by more recent statistical investigations. The need to select from the measurements the appropriate velocity scales makes the Fourier technique ideal for studying correlations between solar atmospheric quantities (Edmonds & Webb, 1972a, b). As well as estimating the power spectra of two fields, X and Y say, we may also estimate the cross spectrum P_{XY}, i.e. the Fourier transform of the cross-covariance function (cf. Eqn. 2.7)

$$C_{XY} = \int_{-\infty}^{\infty}\int X(\xi_1, \xi_2)\, Y(\xi_1 + x_1, \xi_2 + x_2)\, \mathrm{d}\xi_1\, \mathrm{d}\xi_2. \tag{2.14}$$

Since the cross-covariance is again an unnormalized form of the cross-correlation function, it is appropriate to define the quadratic coherence (see Jenkins & Watts, 1969) in a dimensionless manner

$$q^2(k_\perp) = P^2_{XY}(k_\perp)/(P_{XX}(k_\perp)\, P_{YY}(k_\perp)). \tag{2.15}$$

The quantity q is then a correlation coefficient appropriate to the structures with wavenumbers in a band centred on the sample value $k_\perp$. This is equivalent to filtering out structures of both larger and smaller scales and then determining the linear correlation coefficient.

The coherence between the fluctuations at various heights in the solar atmosphere has been studied by, amongst others, Edmonds (1962a), Edmonds, Michard & Servajean (1965) and Edmonds & Webb (1972a, b), but the most extensive study remains that of Canfield & Mehltretter (1973). These authors tabulated various quadratic coherences in the two wavenumber ranges $1.64 < k_\perp < 3.57\ \mathrm{Mm}^{-1}$ ($2''.4 < \Lambda_\perp < 5''.3$) and $3.57 < k_\perp < 5.40\ \mathrm{Mm}^{-1}$ ($1''.6 < \Lambda_\perp < 2''.4$). The coherence between the continuum brightness fluctuation and the Doppler velocity shift in the weakest line was found to be 0.74 and 0.81 in the two ranges. The equivalent correlation coefficients are 0.86 and 0.90. The effect of filtering out both large and small scales is thus to increase the correlation between the brightness and velocity fluctuations in the lower atmosphere. Such values put the convective origin of the granular fluctuations beyond doubt. A note of dissent has been raised by Karpinsky (1979), who claimed a correlation of less than 0.4, but this result must be attributed to a failure to filter out intrusive velocity fields of non-convective origin.

Of equal importance to theoretical modelling of the granulation is the variation of the correlation with height. This was also measured by Canfield & Mehltretter; the relevant approximate heights of the intensity and velocity determinations were calculated by Keil & Canfield (1978). Table 2.12 presents the results for the quadratic coherence in two adjacent wavenumber ranges in each of which the granulation is the dominant scale. The second column gives the total correlation coefficient r between the brightness fluctuations in the continuum and those in the cores of successively stronger lines. The third and fourth columns demonstrate the rise in (quadratic) correlation when the wavenumber is restricted to the bands $1.64 < k_\perp < 3.57\ \mathrm{Mm}^{-1}$ (I) and $3.57 < k_\perp < 5.40\ \mathrm{Mm}^{-1}$ (II). The fifth and sixth columns show the coherence between the continuum brightness fluctuation and the velocity shifts of the cores of successively stronger lines.

A clear picture emerges. On the one hand, intensity structures rapidly lose their identity with height, demonstrated by the drop in correlation and coherence. The small structures become uncorrelated more rapidly than the larger structures (see Kneer, Mattig, Nesis & Werner, 1980; Durrant & Nesis, 1981). The upper photospheric fine structure is almost uncorrelated with the granulation. There is some indication that it may be weakly negatively correlated, that is to say, bright granules tend to turn into dark structures in the upper photosphere. This is evidenced by the change of sign of the correlation coefficient; it cannot be seen in the coherence which is a quadratic quantity. It is the dark structures which are apparent in filtergrams taken in the cores

Table 2.12. *Correlation between brightness and velocity fluctuations I_l and V_l in Fraunhofer lines of increasing strength and the granular brightness fluctuation I_c in the continuum*

Line (nm)	Correlation coefficient, I_c vs I_l	Coherence, I_c vs I_l		Coherence, I_c vs V_l		Mean height (km)	
		I[a]	II[a]	I[a]	II[a]	Intensity	Velocity
517.88	0.845	0.867	0.865	0.739	0.806	40	149
516.46	0.525	0.570	0.499	0.755	0.800	87	178
518.01	0.450	0.447	0.386	0.754	0.801	96	183
515.91	0.019	0.197	0.207	0.770	0.847		
516.54	−0.035	0.321	0.249	0.719	0.734	260	291
516.23	−0.050	0.327	0.285	0.685	0.780		
517.16	−0.190	0.296	0.341	0.282	0.416	447	471

[a] I: $1.64 < k_\perp < 3.57\ \mathrm{Mm}^{-1}$ or $2''.4 < \Lambda_\perp < 5''.3$; II: $3.57 < k_\perp < 5.40\ \mathrm{Mm}^{-1}$ or $1''.6 < \Lambda_\perp < 2''.4$.

of strong lines (see Section 2.4.2). On the other hand, the velocity structures pervade the whole atmosphere almost unchanged. Only in the very strongest line is the correlation between continuum brightness and core velocity significantly lowered. This is confirmed by the large coherences between the Doppler shifts in lines of various strengths included in the earlier investigation of Edmonds *et al.* (1965).

The coherence analysis thus brings into focus the interpretation of the brightness and velocity observations that has emerged in the previous sections. The granular intensity field rapidly disappears but the associated velocity field continues almost unabated with height. The high-quality spectrogram obtained at the Pamir observatory and shown in Fig. 2.16(*b*) demonstrates how the granular shifts can be traced between lines of all strengths, even to the strongest line present, Mg I λ470.3 nm. The correlation between the intensity and velocity fluctuations in the cores of lines of increasing strength rapidly disappears. This provides compelling evidence for the overshoot of convective motions into the upper photosphere and a strong radiative damping of the temperature perturbations. The granulation itself is a direct manifestation of the convective transport of heat but the other structures of the photosphere are not.

The study of spatial correlations is evidently a powerful tool for the interpretation of observations of the solar atmosphere. The study of spatio-temporal correlations is potentially as important. Edmonds *et al.* (1965) studied the correlation between granular brightness and long-period Doppler shifts measured in low photospheric lines and found a significant phase shift, the motion preceding the brightness by some 60° (i.e. a time advance of about 100 s). Musman (1974), averaging over all horizontal scales, reported the advance to be about 60 s.

Edmonds & Webb (1972b) gave a revised estimate of the brightness lag of about 20 s, independent of the horizontal scale of the fluctuation, based on a reanalysis of the data of Edmonds *et al.* (1965). Deubner (1974b), on the other hand, found from higher-quality material that the phase switches from a brightness lag of 50 s at large scales to an advance of 60 s at scales smaller than $2\overset{''}{.}4$.

The significance of these conflicting results is hard to assess. The very existence of a phase difference indicates that we are not dealing with a stationary cellular flow but a time-dependent process in which there is an interplay between the velocity and brightness fluctuations. We shall return to this topic in Section 5.2.4.

2.6 Granulation and magnetic fields

Early magnetograms taken with low spatial resolution appeared to reveal the existence in the photosphere of weak large-scale magnetic fields covering substantial areas of the solar disk; these were frequently present even in quiet regions showing no other sign of activity. Several early workers looked for fine structure in these fields and sought a relationship with the photospheric granulation. Using a conventional spectrograph equipped with a polarizing device Steshenko (1960) placed an upper limit of about 5 mT on the strength of any longitudinal field possibly associated with individual granules. On the other hand, he found that groups of granules coinciding with regions of highest upward velocity were associated with longitudinal fields of the order of 5-6 mT. Semel (1962) found no indication of any strong correlation between the field distribution and the granulation pattern in an area with a mean field strength of about 2.4 mT. Livingston (1968) using a magnetograph with a claimed resolution of 500 km placed an upper limit of 0.2 mT on any field correlated with the granulation pattern. In the network he found fields from 1-10 mT.

Our ideas about the nature of photospheric magnetic fields have changed markedly in recent years (see, for example, the Joint Discussion on the small-scale structure of solar magnetic fields edited by Deubner (1977)). It is now generally accepted that most fields are concentrated into regions with a diameter of $\sim 0''.1$-$0''.3$ - below the resolution limit of present-day telescopes - where the fields are some 100-200 mT. The spatial resolution of 1-2″ generally achieved by spectrographs and magnetographs leads to a smearing of the image which reduces the measured field strength of a single element by a factor of a hundred or so. Thus the fields measured by the earlier authors occupied only a small fraction of the effective instrumental aperture. Indeed, the fields occupy a much smaller area than the typical granule and have no obvious relationship with the granulation pattern as a whole.

Whether or not there exists another weaker and more diffuse component of the photospheric magnetic field is at present uncertain. If such fields exist, their strengths cannot exceed a few mT at the very most; fields of this magnitude could be expected to have little or no influence on the dynamics of the individual granules (see Danielson, 1966).

Within plage regions and at the edges and vertices of supergranulation cells the measured mean field strengths are no more than 10-20 mT. According to our current picture of concentrated magnetic field elements, this can result only if the elements occupy a fraction (10-20 mT)/(100-200 mT), i.e. about 10%, of the surface area. Most of the surface is thus not occupied by magnetic fields and we should not expect the general pattern of the granulation to alter very much. Measurements of magnetic fields of sunspots indicate that the apparent field strength drops very rapidly within about 10″ of the penumbral boundary (Hagyard *et al.*, 1982); the influence of the spot fields on the granulation in their vicinity should be correspondingly limited. Nevertheless, several authors have observed a reduction in the granule diameter or mean spacing in the close neighbourhood of sunspots (cf. Section 2.3.2), which so far has received no theoretical attention.

Limited areas of 'small-scale' granules also occur irregularly in the general field of the granulation, according to Edmonds (1960, 1962b). But these appear to be chance clumpings, perhaps due to the simultaneous fragmentation of an entire 'family' of granules, since they last no longer than a typical granule. They probably have nothing to do with magnetic fields.

In contrast to the generally undisturbed appearance of the granulation in the neighbourhood of well-established sunspots, cases of definite disturbance were first observed by Loughhead & Bray (1961) in the granulation between new and developing pores (see also Miller, 1960). The disturbance took the form of a number of dark lanes running roughly parallel to a line joining the pores. This

Fig. 2.25. Disturbed granulation near a developing active region, photographed with the 76-cm vacuum tower telescope of the Sacramento Peak Observatory. The upper left shows normal granulation, while the lower right shows abnormal granulation peppered with tiny filigree elements. The dark elongated structures between the sunspot and the pores are thought to be produced by emerging magnetic loops. (By courtesy of the Association of Universities for Research in Astronomy, Inc., Sacramento Peak Observatory.)

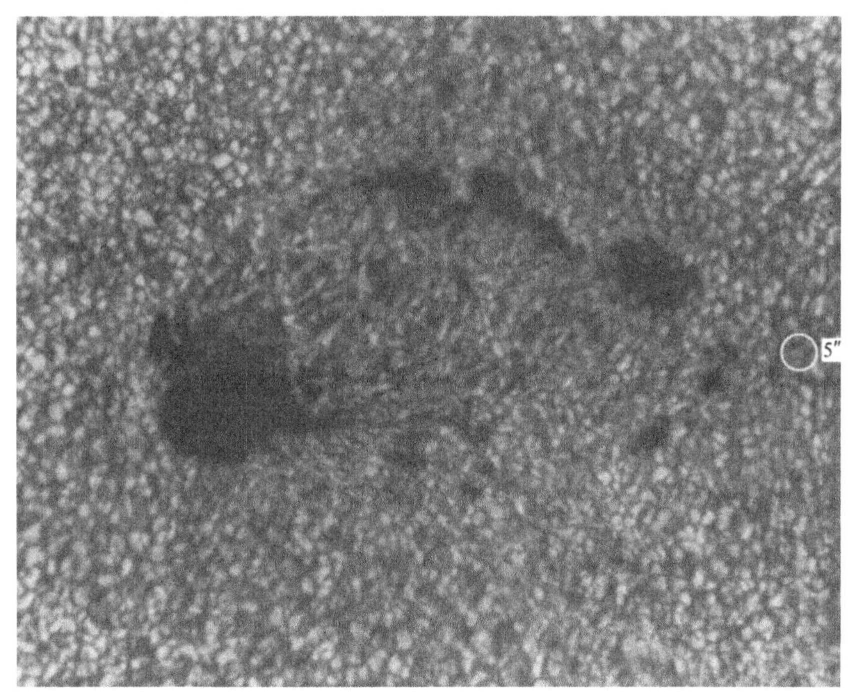

phenomenon is shown in Fig. 2.25. Such observations appear to provide the only evidence so far for a direct interaction between convective flow and strong magnetic fields in the solar atmosphere. It has been observed only rarely and it is thought to occur when magnetic loops are in the process of emerging through the solar surface (Loughhead & Bray, 1961; Cram, 1980).

More common is an optical, rather than a dynamical, change of the granulation in response to activity. Edmonds (1960, 1962b) noticed in several Stratoscope frames areas of abnormal granulation. These areas give the impression of a blurring of the granulation, implying a lowering of the contrast. This effect is not well understood, but an important factor was noted on the basis of filtergrams by Bray *et al.* (1974) and can also be seen in the photographs of Fig. 2.14. The areas of abnormal granulation in the white-light picture, Fig. 2.14(*a*), coincide with regions of enhanced brightness, the photospheric network, in the narrow-band picture, Fig. 2.14(*b*). Many of the bright network elements are cospatial with individual smeared granules.

It is possible that the abnormal granulation may be related to a phenomenon called 'filigree', first described by Dunn & Zirker (1973). The filigree is visible in both wings of the Hα line as an irregular network of very small bright elements or 'crinkles'. It is also faintly visible in the continuum (see the lower right-hand side of Fig. 2.25). Dunn & Zirker found that at first glance there is a strong suggestion that the filigree lies within the intergranular lanes, thus lowering the contrast of the granules and producing a region of abnormal granulation. On the other hand, they found that this is not true everywhere. The possible relationship between crinkles and photospheric magnetic elements has been discussed by several authors (Dunn & Zirker, 1973; Mehltretter, 1974; Simon & Zirker, 1974), but without magnetic observations of much higher spatial resolution no definite conclusion can be drawn.

2.7 The supergranulation

The existence of a large-scale system of horizontal velocities in the upper levels of the photosphere was first discovered by Hart (1954, 1956) at Oxford in the course of a new determination of the Sun's rotational velocity, undertaken with a view to elucidating the origin of certain discrepancies among the results of earlier determinations. She found that the individual motions occurred on an irregular scale ranging from about 25 000 to 85 000 km and persisted for at least several hours; the velocities involved were as high as 0.5 km s^{-1}. However, it was not until 1960, when Leighton at the Mt Wilson Observatory pioneered his new spectroheliographic technique of 'photographing' solar velocities (see Section 2.5.4), that the true extent and significance of the horizontal motions were fully realized. In the following years, thanks to the continued efforts of Leighton and his collaborators (Leighton *et al.*, 1962; Leighton, 1963; Simon, 1964; Simon & Leighton, 1964), quite a detailed picture was built up both of the structure and properties of the horizontal velocity field and of its relationship to other solar phenomena.

The extent and regularity of the system of horizontal motions is immediately apparent from Fig. 2.26. This is a typical Doppler photograph of the Sun, taken

Fig. 2.26. Doppler photograph of the Sun showing the large-scale distribution of horizontal motions constituting the supergranulation (Leighton, 1963). Light and dark areas indicate velocities directed towards and away from the observer respectively (the effect of solar rotation has been largely removed). The velocity field is made up of numerous velocity 'cells', each many thousands of kilometres in diameter, standing out against a grey (stationary) background. Note that in each case the line-of-sight velocity is directed towards the observer on the side of the cell facing towards the centre of the disk and away from the observer on the side facing the limb.

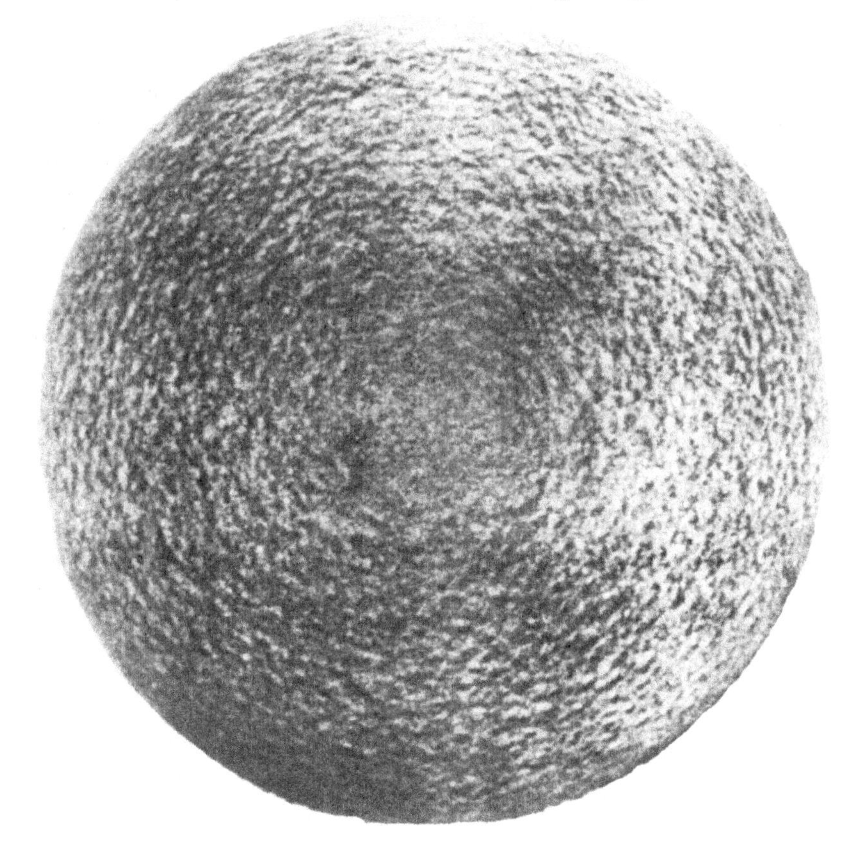

in the line Ca I λ610.3 nm, in which the line-of-sight velocities appear as light or dark shading according to whether they are directed towards or away from the observer and in which the effect of solar rotation has largely been removed. It is evident that the velocity field is made up of numerous velocity 'cells', each many thousands of kilometres in diameter, standing out against a grey (stationary) background. Near the limb the cells show evidence of geometrical foreshortening. Near the centre of the disk they are hardly visible at all, thus indicating that the velocities are predominantly horizontal. There is no detectable difference in the appearance of the velocity pattern between equatorial and polar regions, but it is more difficult to study at latitudes exceeding ±60° (Leighton *et al.*, 1962).

Most of the individual velocity cells visible on Fig. 2.26 display one remarkable regularity: the line-of-sight velocity is directed towards the observer on the side of the cell facing the centre of the disk and away from the observer on the side facing the limb. This is manifested by the fact that the cells appear bright on the centre side and dark on the limb side. This observation may be interpreted as implying that the motion in each cell is directed outwards from the centre towards the boundary.

In their (1962) paper Leighton and his coworkers pointed out that the horizontal velocity field bears a certain resemblance to the ordinary photospheric granulation, although the spatial and temporal scales are, of course, much larger. The velocity field is so well ordered and long-lived that it was highly suggestive of the flow patterns observed in cellular convection (see Section 3.4.3). Leighton *et al.* therefore came to the conclusion that the horizontal velocity field was to be identified with a giant system of convection currents, which they named the 'supergranulation', a term that has since found general acceptance.

Most of the main quantitative parameters pertaining to the supergranulation now seem fairly well established. The available data may be briefly summarized as follows:

(1) *Mean cell size*. Leighton *et al.* (1962) found that at any given time there are roughly 2500 cells on the visible solar hemisphere. From this one can immediately derive an estimate of the mean cell size, or the average distance between the centres of adjacent cells in the velocity pattern (see Section 2.3.3). If we call this distance d_s, the total number of cells must be of the order of $2\pi R_\odot^2/d_s^2$ (where $R_\odot$ is the solar radius) and, equating this to the observed figure of 2500, we obtain

$$d_s \simeq 35\,000 \text{ km}$$

for the mean cell size of the supergranulation pattern. Like the photospheric granulation, the supergranulation exhibits a fairly large spread in the individual cell diameters which, according to the results of an autocorrelation analysis carried out by Simon & Leighton (1964: see Fig. 4), range from 20 000 to 54 000 km. Their mean cell size, 30 000 km, obtained from the FWHM of the autocorrelation function, has been confirmed by Duvall (1980).

Frazier (1970) studied 20 individual cells and determined the mean distance between opposite vertices, defined as the meeting-points of several cells. The 'diameter' found in this relatively small sample was also 30 000 km.

(2) *Lifetime*. The observations of Hart (1956) and of Leighton *et al.* (1962) show that the individual velocity cells have lifetimes of at least several hours. More precise measurements are difficult to make since the cells, being virtually invisible in the central region of the disk, have to be studied at heliocentric angles where the geometrical foreshortening is rather severe.

Worden & Simon (1976) found that after nine hours the correlation between two supergranulation photographs had dropped by only some 15%; they concluded that the lifetime must be of the order of a day or so. Observations extending over two days allowed Duvall (1980) to improve the estimate to 42 ± 5 hr.

Simon & Leighton (1964) showed that a strong spatial correspondence exists between the cell boundaries and the network of bright calcium K-line elements in the overlying chromosphere and found that the latter last for periods of the order of 20 hr, in essential agreement with the result obtained earlier by Macris (1962). Since it is plausible that the lifetime of the motions which are believed to collect and maintain the network is at least as long as that of the network itself, these values provide lower limits to the lifetime of the supergranulation.

(3) *Magnitude of the horizontal velocities*. The measurements of a number of observers (Hart, 1956; Evans & Michard, 1962b; Simon & Leighton, 1964) agree in giving values of the order of 0.3-0.5 km s^{-1} for the average peak horizontal velocities within the individual cells. The magnitude of this streaming velocity appears to decrease with height in the atmosphere (Hart, 1956; Deubner, 1971; Edmonds & Webb, 1972b). There is some evidence that the motion carries the individual granules with it (Simon, 1967).

(4) *Existence of vertical velocities*. If the interpretation of the horizontal velocity field in terms of a system of convection cells is correct, then one should expect to find evidence of vertical velocities at both the centres and boundaries of the individual cells. And, indeed, on a few of their best Doppler photographs Simon & Leighton (1964) detected in

the central region of the disk a *very faint* pattern of rising velocities at the centre of each cell and a network of descending motions at the cell boundaries. As Simon (1964) showed, the difficulty in observing the vertical components of the large-scale system is due to the presence of the prominent and quite unrelated regime of vertical oscillatory velocities which we have described in Section 2.5.4. Nevertheless, by applying an appropriate method of time-smoothing to a series of spectrograms taken at the Sacramento Peak Observatory, Simon was able to obtain independent confirmation of the existence of downward velocities of the order of 0.1-0.2 km s^{-1} at the boundaries of the supergranules. This figure has since been confirmed by several other authors.

A more detailed picture is given by Frazier (1970), who measured average cross-sections of supergranulation cells and found that the downflows at the vertices where several cells meet are much more prominent than those at the 'walls' joining such vertices. The downflow is strictly associated with the network elements and strong magnetic fields at the boundaries of the cells. An intriguing question which has yet to be answered is how much of the flow is confined within the magnetic elements gathered at the cell boundary and how much takes place outside them (cf. Giovanelli & Slaughter, 1978). It is noteworthy that a corresponding upflow in the centre of the cell, uninfluenced by magnetic fields, is much more difficult to establish. Frazier (1970) concluded that the magnitude of the upflow was about half that of the downflow; Worden & Simon (1976) quote a value of 0.05 km s^{-1}.

By averaging out the small-scale granular fluctuations Beckers (1968b) and several subsequent workers found that the mean photospheric brightness is slightly greater at the boundaries of the supergranules than at their centres. The correlation between brightness and vertical motion is thus opposite to that expected of a convective phenomenon. On the other hand, at a wavelength of 1.64 μm, where the solar opacity is at a minimum and the deepest atmospheric layers are revealed, Worden (1975) found that the centres appeared brighter than the boundaries. The *possible* identification of the supergranulation as a *convective* phenomenon is discussed in Section 5.4.

Readers interested in the correlation between the supergranulation and other solar phenomena are referred to the exhaustive discussion in Bray & Loughhead (1974).

November, Toomre, Gebbie & Simon (1981) have recently claimed to have detected a cellular velocity pattern in the upper photosphere that has a scale of 5-10 Mm (7-14″) and a lifetime of at least two hours. The rms vertical velocity is reported to be 60 m s^{-1}, comparable to the vertical velocities in the supergranulation but much weaker than the granular velocities. The relationship of

Table 2.13. *Summary of properties of photospheric granules and other photospheric and sunspot structures*

Feature	Diameter	Mean cell size	Lifetime	Velocity ($km\ s^{-1}$)	Reference[a]
Photospheric granules	1–2″	1″.9	~16 min	~1 (rms, vertical) ~2 (rms, horizontal)	2.3.2, 2.3.3, 2.3.6, 2.5.4
Facular points	<0″.5		12–35 min	~2 (downflow)	Mehltretter (1974); Muller (1977, 1981)
Facular granules (conglomerates)	1–2″		2 hr		Bray & Loughhead (1961)
Umbral dots	≲0″.5	0″.9–1″.3	15–30 min	~3 (upflow)	**3.6**; Kneer (1973); Loughhead, Bray & Tappere (1979); Koutchmy & Adjabshirzadeh (1981); Moore (1981)
Penumbral bright grains (aligned in filaments)	≲0″.5 × 2″.0		40–180 min	0.0–0.5 (horizontal)	**3.5.4**, **3.5.6**; Muller (1973, 1976); Moore (1981)
Supergranulation	30–70″	48″	≳1 day	0.3–0.5 (horizontal)	2.7

[a] Section numbers in bold type refer to Bray & Loughhead's (1964) monograph, those in ordinary type to this text. (Note: 1″ = 725 km).

this flow field, which they propose to call 'mesogranulation', to the other phenomena discussed in this book remains unclear. We should, however, draw attention to the similarity of scale and lifetime between the 'mesogranulation' and the families of granules mentioned in Sections 2.3.7. and 2.5.2. In both cases, it is clear that motions of presumably convective origin in the solar photosphere exhibit collective properties more extended in time and space than the commonly accepted picture of the granulation would imply. Observations are continuing to give new impetus to the development of our understanding of solar convective phenomena.

2.8 Summary of data

To conclude this chapter, we summarize in Table 2.13 the known quantitative data concerning the main physical properties of the photospheric granules and supergranules as well as, for comparison, the corresponding data foı other photospheric and sunspot structures. This table provides a useful résumé of the similarities and dissimilarities among the various features, but for more comprehensive information the reader should refer to the appropriate sections of this book, Bray & Loughhead's (1964) monograph, or the references given in the last column.

Additional notes

The consistency of Fried parameter values derived from turbulence measurements, wave-front distortion, and image motion monitors has been demonstrated by J. Borgnino, G. Ceppatelli, G. Ricort & A. Righini (1982, *Astronomy and Astrophysics*, **107**, 333-7) (Section 2.2.2).

Important but conflicting contributions to the question of the height dependence of the vertical granular velocity have been made by S. L. Keil & F. H. Yackovich (1981, *Solar Physics*, **69**, 213-21) and M. Bässgen & F.-L. Deubner (1982, *Astronomy and Astrophysics*, **111**, L1-3) (Sections 2.4.2 and 2.5.4).

3

AN INTRODUCTION TO THE THEORY OF CONVECTION

3.1 Introduction

As we have seen in Chapter 2, the photospheric granulation displays a clearly defined *cellular* pattern with a relatively narrow distribution of cell sizes, the mean distance between the centres of adjacent granules being about 1400 km (Section 2.3.3). This basic observational fact provides the chief reason for regarding the granulation as a convective phenomenon: the only physical mechanism known to produce such a pattern in a fluid is buoyancy-driven convection, apart from surface tension and other agencies known to be absent. This conclusion is supported by the existence of a strong correlation both in sense and in magnitude between brightness and velocity in the photosphere, the bright features showing violet Doppler shifts (upward-moving material) and the dark features showing redshifts (Section 2.5.5).

Moreover, convective instability is a necessary consequence of the presence of a zone of partial ionization of the dominant element, hydrogen, a few hundred kilometres below the solar surface. An elementary volume of gas moving upwards through such a region of rapidly decreasing ionization liberates ionization energy which, in turn, is converted to thermal energy. Thus the buoyancy of the elementary volume is increased and it continues its upward journey, provided the forces opposing its motion are sufficiently small.

Our chief aim in this and the following chapter is to provide a proper theoretical basis for interpreting the observed properties of the photospheric granulation and supergranulation, in so far as the current state of the theory of convection permits. In pursuing this aim, we have not hesitated to include not only discussions of astrophysical convection in general and the solar convection zone in particular, but also an account of modern work on the theory of convection *per se*.

The subject matter is inherently difficult and complex, but we have tried to develop the account in such a way as to make the subject accessible both to solar physicists and astrophysicists with no expert knowledge of modern fluid mechanics and to theoretical hydrodynamicists with no previous experience of solar physics.

We begin (Section 3.2) with the general equations of hydrodynamics and then consider the ways in which they may be simplified in the context of astrophysical convection. Since the observed convective velocities are low enough for pressure-induced effects (acoustic waves) to play a negligible role, we may suppress certain terms and thus obtain the so-called anelastic approximation (Section 3.2.3). This approximation provides a consistent description of convective motions in stratified fluids, such as those in the Sun and stars. Stratified fluids are almost impossible to realize in the laboratory and most experiments are adequately described by uniform or incompressible models. Under such circumstances a further simplification of the equations is possible, leading to the Boussinesq approximation. Because of the mathematical simplicity and the opportunity for experimental investigation most studies of convection have been based on the Boussinesq approximation. It does not, however, provide a good model for solar convection.

The onset of convection is discussed from a theoretical point of view in Section 3.3. Consideration of the marginal stability of a horizontal layer of fluid heated from below – the Rayleigh problem – introduces the Rayleigh number, $\mathscr{R}a$. The magnitude of $\mathscr{R}a$, a dimensionless parameter which depends on the temperature gradient in the fluid and on its physical properties, determines whether or not the fluid is convectively unstable. The condition for instability can be written $\mathscr{R}a > \mathscr{R}a_c$, where $\mathscr{R}a_c$ is approximately equal to 1000, the exact value depending on the system and its boundary conditions. However, $\mathscr{R}a$ actually has a far wider significance because both observation and theory show that, for fluids which are beyond the state of marginal instability, the character of the convection depends mainly on the magnitude of $\mathscr{R}a / \mathscr{R}a_c$.

Information on the behaviour of convecting fluids in the laboratory is limited to the Boussinesq case (Section 3.4). Initially, steady well-ordered patterns of cells or rolls are observed but, as we move away from the state of marginal stability, the steady motions become disrupted by unsteady swirling in the horizontal plane. The onset of these time-dependent phenomena is governed by another dimensionless parameter, the Prandtl number $\mathscr{P}\!r$, which depends only on the fluid properties. The lower the Prandtl number the more rapidly the swirls build up. In fluids with moderate values of $\mathscr{P}\!r$ the swirls first produce steady oscillations, which then give way at higher $\mathscr{R}a/\mathscr{R}a_c$ to a whole range of eddy motions – a process that leads to turbulence:

> 'Big whirls have little whirls, which feed on their velocity,
> Little whirls have smaller whirls, and so on, to viscosity.'

The properties of turbulent convection are briefly reviewed in Section 3.4.4.

In Section 3.4.5 we look at some properties of the boundary between a convectively unstable layer and an overlying stable layer. A convenient system of this kind is the so-called ice-water experiment which, moreover, may be highly turbulent. The turbulent boundary layer provides a valid prototype for the granulation. The experimental findings are very striking: they demonstrate clear structuring even within highly turbulent convective regions. We conclude with a discussion of the non-linear processes responsible for these structures and of the appearance of internal waves in the stably stratified layer.

3.2 Basic equations and their simplification

3.2.1 Notation and units

A spherical coordinate system is most appropriate for studies of stellar structure, but we shall be concerned mainly with the outermost layers of the Sun and its atmosphere. Since the thickness of these layers is small compared to the solar radius, we may neglect the effect of curvature and introduce a Cartesian coordinate system Ox_i $(i = 1, 2, 3)$. We choose the x_3-axis to be directed upwards, so the gravitational acceleration may be written as

$$g_i = -\delta_{i3} g.$$

The zero point of the vertical axis will generally be taken at the upper boundary of the layer with convectively unstable stratification. In the Sun this point coincides, to within a few kilometres, with the level $\tau = 1$ at $\lambda 500$ nm (Section 5.2.2), which we define to be the 'base' of the atmosphere. Use will be made of both Cartesian tensor and vector notation according to convenience.

The principal symbols employed throughout the book are defined in Table 3.1; SI units are used throughout.

3.2.2 Hydrodynamic equations

The photosphere is composed largely of unionized hydrogen gas at temperatures between 4000 and 7000 K and at pressures ranging from 10^2 to $10^4\,\mathrm{N\,m^{-2}}$, and can be treated as a one-fluid plasma. Furthermore, the absence of significant magnetic fields in the normal granulation (Section 2.6) means that we can interpret photospheric motions within the framework of ordinary fluid mechanics. The equations may be found under several guises (see, for example, Landau & Lifshitz, 1959; Yih, 1969; Truesdell & Muncaster, 1980) but, for our purpose, may be conveniently stated as follows:

(1) *Mass conservation or continuity*

$$\frac{\partial \rho}{\partial t} + \frac{\partial}{\partial x_j}(\rho v_j) = 0. \tag{3.1a}$$

Table 3.1. *Principal symbols*[a]

A	area
A_{ij}	Einstein coefficient
B	Planck function
B_{ij}	Einstein coefficient
C	contrast; selfinteraction coefficient
C_{ij}	collision cross-section
C_p	specific heat at constant pressure per unit mass
C_{XX}	covariance function
C_{XY}	cross-covariance function
c	speed of light (2.9979×10^8 m s^{-1})
c_s	speed of sound
D	diameter
d	dimension
E	kinetic energy
e	internal energy per unit mass
$\mathbf{F}$	heat flux vector
f	horizontal planform function
G	gravitational constant (6.67×10^{-11} N m^2 kg^{-2})
g	gravitational acceleration at solar surface (274 m s^{-2})
H	scale height
h	enthalpy per unit mass; Planck's constant (6.625×10^{-34} J s)
I	intensity (brightness)
J	mean intensity
j_λ	mass emission coefficient
K	coefficient of conductivity
$\mathbf{k}$	wave vector
k	Boltzmann's constant (1.380×10^{-23} J K^{-1})
k_λ	mass absorption coefficient
L	turbulent drag length
l	mixing length
$M_\odot$	solar mass (1.99×10^{30} kg)
m	polytropic index
N	number
n	growth rate
n_i	number density
P_{XX}	power spectrum (two-dimensional)
P_{XY}	cross spectrum (two-dimensional)
p	pressure
q	coherence
R	radius; gas constant (8.317 J K^{-1})
$R_\odot$	solar radius (6.96×10^8 m)
R_*	gas constant per unit mean molecular weight
r	correlation coefficient
S	source function
s	path length; entropy per unit mass
T	temperature
$\mathbf{v}$	velocity vector

W	weighting function
$\mathbf{x}$	position vector
X_0	initial quantity
$\bar{X}$	mean quantity
X'	fluctuation about mean or initial value
$\hat{X}$	typical value of a quantity X
$\langle X \rangle$	root mean square value of a quantity X
$\tilde{X}$	quantity X in non-dimensional form
$X_\perp$	horizontal component of vector **X**
$\mathcal{M}$	Mach number
$\mathcal{N}u$	Nusselt number
$\mathcal{P}e$	Péclet number
$\mathcal{P}r$	Prandtl number
$\mathcal{R}a$	Rayleigh number
$\mathcal{R}e$	Reynolds number
α	compressibility; ratio of mixing length to scale height
β	superadiabatic gradient
Γ	aspect ratio
γ	ratio of specific heats
δ	volumetric expansion coefficient
δ_{ij}	Kronecker delta tensor
ϵ	energy generation rate
ϵ'	thermal coupling parameter
ϵ_{ijk}	alternating tensor
ζ	bulk viscosity
η	dynamic viscosity
Θ	eddy growth factor
θ	heliocentric angle
κ	opacity
Λ	spatial wavelength
λ	spectral wavelength
μ	$\cos\theta$
ν	kinematic viscosity
$\boldsymbol{\xi}$	position vector
ρ	mass density
σ	Stefan–Boltzmann constant ($5.669 \times 10^{-8}\,\mathrm{W\,m^{-2}\,K^{-4}}$)
σ_{ij}	viscous stress tensor
τ	optical depth
Φ	eddy geometric factor
χ	thermal conductivity
Ψ	eddy geometric factor
Ω	solid angle
ω	temporal frequency
ϖ	statistical weight
∇	logarithmic gradient

[a] SI units are used throughout.

The local rate of change of density is balanced by the divergence of the mass flux, i.e. the net rate at which mass flows into the infinitesimal volume. If the second term is expanded, another standard form results

$$\mathrm{D}\rho/\mathrm{D}t \equiv \partial\rho/\partial t + v_j\, \partial\rho/\partial x_j = -\rho\, \partial v_j/\partial x_j. \tag{3.1b}$$

The operator on the left-hand side, $\mathrm{D}/\mathrm{D}t$, is the material derivative. It accounts for both the local change with time and the change due to transport of the fluid element with velocity $\mathbf{v}$ through a region of spatially varying properties. The second component is sometimes called the 'convective' term but in the present context the less confusing word 'advective' is preferable.

(2) *Momentum conservation*

$$\frac{\partial}{\partial t}(\rho v_i) + \frac{\partial}{\partial x_j}(\rho v_i v_j) = \rho g_i - \frac{\partial p}{\partial x_i} + \frac{\partial \sigma_{ij}}{\partial x_j}. \tag{3.2a}$$

This equation states that the local rate of change of momentum and the divergence of the momentum flux are maintained by the gravitational body force and by the pressure and viscous stress forces. It can be rewritten, using Eqn. 3.1, as an equation for the fluid acceleration

$$\rho \mathrm{D}v_i/\mathrm{D}t \equiv \rho\, \partial v_i/\partial t + \rho v_j\, \partial v_i/\partial x_j = \rho g_i - \partial p/\partial x_i + \partial\sigma_{ij}/\partial x_j. \tag{3.2b}$$

The advective term in this equation is sometimes called the 'inertial' term.

Under conditions obtaining in the solar atmosphere the gas behaves as a Newtonian fluid and the stress tensor σ_{ij} is linearly related to the local shear through the dynamic and bulk viscosities (η, ζ respectively):

$$\sigma_{ij} = \eta(\partial v_i/\partial x_j + \partial v_j/\partial x_i) + (\zeta - 2\eta/3)\, \delta_{ij}\, \partial v_k/\partial x_k. \tag{3.3}$$

Furthermore, the atomic relaxation processes are sufficiently rapid to allow us to neglect ζ (Landau & Lifshitz, 1959: p. 304).

(3) *Energy conservation*

Energy conservation can be expressed in many equivalent ways. The fundamental quantity is usually taken to be the internal energy per unit mass e. This is just the heat content at constant volume and obeys a conservation equation that can be manipulated in a manner similar to Eqn. 3.2 to yield

$$\rho\frac{\mathrm{D}e}{\mathrm{D}t} \equiv \rho\frac{\partial e}{\partial t} + \rho v_j \frac{\partial e}{\partial x_j} = -\frac{\partial F_j}{\partial x_j} - p\frac{\partial v_j}{\partial x_j} + \sigma_{jk}\frac{\partial v_j}{\partial x_k}. \tag{3.4a}$$

On the right-hand side the successive terms represent the contributions from the net heat flux, the work done by the pressure forces and the viscous heating. Eqn. 3.4a can be used in combination with well-known thermodynamic relations to generate alternative formulations.

Defining the entropy per unit mass s by means of the relation

$$T\mathrm{D}s/\mathrm{D}t = \mathrm{D}e/\mathrm{D}t + p\mathrm{D}(1/\rho)/\mathrm{D}t, \tag{3.5}$$

(see Morse, 1965) the energy equation takes its simplest form,

$$\rho T\mathrm{D}s/\mathrm{D}t = -\,\partial F_j/\partial x_j + \sigma_{jk}\,\partial v_j/\partial x_k. \tag{3.4b}$$

In ideal fluids dissipation is negligible, so the viscosity and heat flux gradient can be set to zero; entropy is then conserved. We are not in this fortunate position. The flows that interest us are dissipative and compressible. The fluctuations in the flow, as we shall see, take place more often at constant pressure than at constant volume, so the natural thermodynamic parameter is the heat content at constant pressure, the enthalpy per unit mass h, defined by the relation

$$\rho\mathrm{D}h/\mathrm{D}t = \rho T\mathrm{D}s/\mathrm{D}t + \mathrm{D}p/\mathrm{D}t. \tag{3.6}$$

Substitution in Eqn. 3.4*b* yields

$$\rho\mathrm{D}h/\mathrm{D}t - \mathrm{D}p/\mathrm{D}t = -\,\partial F_j/\partial x_j + \sigma_{jk}\,\partial v_j/\partial x_k. \tag{3.4c}$$

In actual experiments, however, neither entropy nor enthalpy is measured, but temperature; the equations may be recast in terms of this parameter using the further relatio..

$$T\mathrm{D}s/\mathrm{D}t = C_p\mathrm{D}T/\mathrm{D}t - (\delta/\rho)\,\mathrm{D}p/\mathrm{D}t. \tag{3.7}$$

Here C_p is the specific heat at constant pressure and δ is the volumetric expansion coefficient $-(T/\rho)(\partial\rho/\partial T)_p$.

We then obtain a closed set of equations in terms of the independent variables $\rho, \mathbf{v}, T$ in which the energy equation appears as

$$\rho C_p\mathrm{D}T/\mathrm{D}t - \delta\,\mathrm{D}p/\mathrm{D}t = -\,\partial F_j/\partial x_j + \sigma_{jk}\,\partial v_j/\partial x_k. \tag{3.4d}$$

These forms all describe only part of the total energy balance. They may be combined with Eqn. 3.2 to yield the conservation of total energy, which is expressed by

$$\frac{\partial}{\partial t}(\rho e + \tfrac{1}{2}\rho v^2) + \frac{\partial}{\partial x_j}(F_j + \rho h v_j + \tfrac{1}{2}\rho v^2 v_j - v_i\sigma_{ij}) = \rho v_i g_i. \tag{3.8}$$

In this form the net rate of change of total energy due to the local time changes of internal and kinetic energy and the divergences of the fluxes of heat, enthalpy, kinetic energy and viscous stresses are balanced against the change of potential energy, i.e. the rate of working against gravity.

In the case of thermal conduction or radiative energy diffusion the heat flux takes the simple form

$$F_i = -\,K\,\partial T/\partial x_i, \tag{3.9}$$

where K is the coefficient of conductivity or radiative diffusivity. This is the form appropriate to most laboratory experiments and to the interior of the Sun.

In the solar atmosphere a more general treatment of radiative transport is necessary; it will be discussed in Section 4.4.3.

Whichever set of equations is chosen, it must be closed by the equation of state for a perfect gas

$$p = R_* \rho T = (R/\bar{\mu})\,\rho T, \tag{3.10}$$

R being the gas constant and $\bar{\mu}$ the mean molecular weight, and by expressions for the functional dependence of the fluid parameters (C_p, δ, η, R_*) on whichever independent variables are chosen. These expressions must allow for ionization in a gaseous mixture; suitable formulae are given by Vardya (1965) and Cox & Giuli (1968: Section 9.18).

Despite their deceptively simple appearance, these equations are formidably complicated. They describe a wealth of physical processes which are coupled through non-linear terms; these make the equations mathematically intractable. A simultaneous consideration of all the resulting effects may ultimately be necessary, but insight can be gained by placing certain restrictions on the type of flow. The most instructive special cases are those in which some processes are coupled only weakly to the others, and thus allow themselves to be isolated.

Just such a simplification can be achieved by noting that the convective velocities in the granulation are small compared with the local sound speed (see Section 3.2.3 below). The general hydrodynamic description allows both buoyancy-driven flows (convection) and pressure-driven motions such as acoustic waves whose phase speeds are greater than or equal to the local speed of sound. If the convective speeds are much smaller than the sound speed there will be a mismatch in either the time scales or length scales, or both, and the coupling between the two forms of motion will be small (see Section 5.5.3). By restricting attention to only those terms with scales appropriate to convection, the waves can be filtered out. The weak coupling cannot be ignored when we wish to explore the generation of acoustic waves by convective motions but, as far as the general description of the convective dynamics is concerned, we can safely neglect it. This filtering forms the basis of the anelastic and Boussinesq approximations.

3.2.3 *The anelastic and Boussinesq equations*

In order to develop these equations, the distinguishing features of our system should be recalled. The observations described in Chapter 2, Sections 2.4.2 and 2.5.4, reveal that the typical granular velocity is less than 1.5 km s^{-1}. The motions are thus subsonic, the speed of sound being about 6 km s^{-1} in the photosphere. Also the continuum intensity fluctuations are no more than 15%. These can be roughly converted into a temperature fluctuation using the

Eddington-Barbier relationship and the Wien approximation (see Section 2.5.3), which yield

$$I'_\lambda/\bar{I}_\lambda \sim B'_\lambda/\bar{B}_\lambda \sim -5.7\,T'/\bar{T}. \tag{3.11}$$

So we see that the fluctuations in the thermodynamic parameters are likely to be correspondingly small. This conclusion will be substantiated by the more rigorous analysis described in Chapter 5. It suggests that the convective motion can be regarded as the perturbation of some notional mean time-independent state. There are several possible ways of defining this background state, leading to different decompositions of the physical quantities into steady and fluctuating components. For the study of the onset of convection it is logical to suppose the time-independent reference system to be an initially static system which is held in equilibrium before release. Then all equations can be rewritten by expanding each quantity as

$$X = X_0(\mathbf{x}) + X'(\mathbf{x}, t), \tag{3.12}$$

where X' is the fluctuating part.

This decomposition is not so advantageous when the convection is well developed and the average value of any quantity may be quite different from its initial value. Then we should isolate the stationary mean component - defined, in theory, as a time or ensemble average but, in practice, much more often as a horizontal spatial average. Because the effects of the motion are then felt in the mean state, the equations tend to be more cumbersome algebraically and will not be given explicitly here; the reader may consult the sources mentioned below.

We shall impose the simplifying conditions:

(1) the initial or mean state varies only in the direction of gravity;
(2) the atmosphere can be taken as plane parallel, thus excluding consideration of the extreme limb;
(3) the gravitational acceleration can be assumed to be constant;
(4) there are no mean flows in the atmosphere;
(5) the atmosphere does not rotate so that neither centrifugal nor Coriolis forces are called into play.

All these conditions are met to a good degree of approximation in the solar atmosphere. The only point calling for comment is the assumption of no rotation. The time scales of the granular and supergranular flows are short compared to the period of rotation of the Sun (~26 days at the equator) and thus their motions are disrupted before the rotational effects can be felt significantly. The observed uniformity of the granulation and supergranulation patterns from equator to pole over the surface of the Sun indeed supports the contention (see Sections

2.3.2 and 2.7.). If flows *longer-lived* than the supergranulation were present in the convective regions, rotational effects would certainly be apparent (see Section 5.4) but we shall ignore such hypothetical questions here.

Let us now return to subsonic flows. Gough (1969) noted that under these circumstances acoustic waves can rapidly and effectively neutralize any pressure imbalance produced by a developing disturbance, so that the (fractional) density and temperature perturbations, which are related through the equation of state, must be comparable. When buoyancy is the main factor driving the vertical motions, a characteristic vertical velocity can be estimated from Eqn. 3.2 as

$$\hat{v}_3'^2 \sim g d_3 \hat{\rho}'/\rho_0 \sim g d_3 \hat{T}'/T_0, \tag{3.13}$$

where d_3 is the vertical dimension of the system.

If we recall that the sound speed c_s is given by

$$c_s^2 \sim p_0/\rho_0 \sim gH, \tag{3.14}$$

where H is the scale height, this relation can be rewritten as

$$\hat{v}_3'^2/c_s^2 = \mathcal{M}^2 \sim (d_3/H)(\hat{T}'/T_0). \tag{3.15}$$

The ratio of flow speed to sound speed, $\mathcal{M}$, is the Mach number of the flow, the first of a series of dimensionless numbers which characterize the system. In order to produce a circulation, these buoyancy-induced vertical motions must be converted into horizontal motions. This requires a fluctuating pressure doing sufficient work to destroy the vertical kinetic energy, which can be estimated from Eqn. 3.4*c* as

$$\hat{p}' \sim \rho_0 \hat{v}_3'^2. \tag{3.16}$$

Hence the relative pressure fluctuation is

$$\hat{p}'/p_0 \sim \mathcal{M}^2. \tag{3.17}$$

If the extent of the system does not encompass too many scale heights, the quadratic velocity and pressure fluctuations as well as the temperature and density fluctuations are small. The equations can then be consistently expanded in terms of the relative fluctuation $\hat{T}'/T_0$. Gough (1969) derived the equations correct to first order in this approximation according to the following prescription:

(1) the equations are linearized in terms of the fluctuating small quantities. The velocity is not regarded as small (see Eqn. 3.16) and non-linear terms involving it are therefore retained;
(2) the time derivative of the density fluctuation in the continuity equation is dropped. This is the crucial step which suppresses the production of acoustic waves.

Eqns. 3.1-3.5, separated into initial and fluctuating parts by such means, become the *anelastic equations*

$$\frac{dp_0}{dx_3} = -g\rho_0 \tag{3.18}$$

$$\frac{dF_{03}}{dx_3} = 0 \tag{3.19}$$

$$p_0 = R_*\rho_0 T_0 \tag{3.20}$$

$$\rho_0 = \rho(p_0, T_0), \text{ etc.} \tag{3.21}$$

$$\beta = -\frac{1}{C_{p0}}\left(\frac{dh_0}{dx_3} - \frac{1}{\rho_0}\frac{dp_0}{dx_3}\right) = -\left(\frac{dT_0}{dx_3} - \frac{\delta_0}{\rho_0 C_{p0}}\frac{dp_0}{dx_3}\right) = -\frac{T_0}{C_{p0}}\frac{ds_0}{dx_3} \tag{3.22}$$

$$\frac{\partial}{\partial x_k}(\rho_0 v_k') = 0 \tag{3.23}$$

$$\frac{\partial}{\partial t}(\rho_0 v_i') + \rho_0 v_k' \frac{\partial v_i'}{\partial x_k} = -\frac{\partial p'}{\partial x_i} - g\rho'\delta_{i3} + \frac{\partial \sigma_{ik}'}{\partial x_k} \tag{3.24}$$

$$\rho_0 C_{p0}\frac{\partial T'}{\partial t} - \delta_0 \frac{\partial p'}{\partial t} - C_{p0}\beta\rho_0 v_3' + v_k'\left(\rho_0\frac{\partial h'}{\partial x_k} - \frac{\partial p'}{\partial x_k}\right) - g\rho' v_3'$$

$$= \sigma_{ik}'\frac{\partial v_i'}{\partial x_k} - \frac{\partial F_k'}{\partial x_k} \tag{3.25}$$

$$\frac{\rho'}{\rho_0} = \alpha_0\frac{p'}{p_0} - \delta_0\frac{T'}{T_0}, \text{ etc.} \tag{3.26}$$

$$\sigma_{ik}' = \eta_0\left(\frac{\partial v_i'}{\partial x_k} + \frac{\partial v_k'}{\partial x_i} - \tfrac{2}{3}\delta_{ik}\frac{\partial v_j'}{\partial x_j}\right). \tag{3.27}$$

We have introduced here the compressibility $\alpha = (p/\rho)(\partial\rho/\partial p)_T$. Similar coefficients arise when we evaluate the fluctuating enthalpy gradient, noting that $h = h(T,p)$. The quantity β measures the entropy gradient in the undisturbed fluid. It is known as the superadiabatic gradient since a system in which β vanishes has constant entropy and is thus adiabatically stratified.

The initial system has a relatively simple mathematical structure in the case of a fluid in which C_p, δ, η, R_* are constant and the heat transfer is described by a conduction equation (3.9) with constant conductivity coefficient. Then the hydrostatic and conductive equilibrium equations (3.18, 3.19) are satisfied by the polytropic solution

$$T_0 = b(-x_3), \quad \rho_0 = P(-x_3)^m/(R_* b), \quad p_0 = P(-x_3)^{m+1}, \tag{3.28}$$

where b, P are constants of integration. Note that the temperature is always a linear function of depth ($-x_3$).

This is the simplest model of a heat-transporting stratified fluid. It has an honourable history, the solution in spherically symmetric geometry providing the archetypal stellar model (see Chandrasekhar, 1939: Chapter 4). Polytropes are distinguished by a functional relationship between the thermodynamic parameters of the form

$$p \propto \rho^{1+1/m}. \tag{3.29}$$

The constant m is known as the polytropic index. In our solution for plane parallel geometry it is given by

$$m = (g/R_* b) - 1. \tag{3.30}$$

The larger the polytropic index the more rapidly the density and pressure increase relative to the temperature or, in other words, the stronger the stratification.

The anelastic equations are the simplest that can be used to describe convection in stratified fluids. Although many terms appear in these equations, they make *explicit* only a few of those *implicit* in the full equations; despite appearances the anelastic equations are easier to solve than the full ones.

A noteworthy further simplification arises if the vertical size of the system is very small compared to the scale height ($d_3 \ll H$), as is normally the case in the laboratory. Then the variation of the background density can be ignored, and the fluctuations in density and pressure can be neglected except in those terms which provide the buoyancy for the vertical motion and the pressure gradient for the horizontal motion. The resulting equations represent the *Boussinesq approximation* (Chandrasekhar, 1961; Gough, 1969). Many authors require a Boussinesq fluid to have, in addition, uniform properties, in which case the Boussinesq equations take the form

$$\mathrm{d}F_{03}/\mathrm{d}x_3 = 0 \tag{3.31}$$

$$\beta = -\,\mathrm{d}T_0/\mathrm{d}x_3 \tag{3.32}$$

$$\partial v'_k/\partial x_k = 0 \tag{3.33}$$

$$\begin{aligned}\partial v'_i/\partial t + v'_k\,\partial v'_i/\partial x_k = {} & -(\partial p'/\partial x_i)/\rho_0 - g\delta_{i3}\rho'/\rho_0 \\ & + \eta(\partial^2 v'_i/\partial x_k^2)/\rho_0\end{aligned} \tag{3.34}$$

$$\partial T'/\partial t + v'_k\,\partial T'/\partial x_k - \beta v'_3 = -(\partial F'_i/\partial x_i)/(C_p\rho_0) \tag{3.35}$$

$$\rho' = -\rho_0\delta T'/T_0. \tag{3.36}$$

Viscous heating disappears from the energy equation in this formulation but, strictly speaking, the viscous forces should be retained in the momentum equation. With a general body force present the momentum equation (3.34) becomes the familiar Navier–Stokes equation.

In the case of conductive heat transport with a constant conduction coefficient, the static structure again possesses a simple linear temperature gradient

$$dT_0/dx_3 = b = -\beta. \tag{3.37}$$

The actual temperature gradient and the superadiabatic gradient are identical in this case as there is no stratification. The Boussinesq equations can then be rewritten in the more standard form

$$\partial v_k'/\partial x_k = 0 \tag{3.38}$$

$$\partial v_i'/\partial t + v_k' \, \partial v_i'/\partial x_k = -(\partial p'/\partial x_i)/\rho_0 + g\delta T'\delta_{i3}/T_0 + \nu \partial^2 v_i'/\partial x_k^2 \tag{3.39}$$

$$\partial T'/\partial t + v_k' \, \partial T'/\partial x_k - \beta v_3' = \chi \partial^2 T'/\partial x_k^2. \tag{3.40}$$

We have introduced here the kinematic viscosity

$$\nu = \eta/\rho_0, \tag{3.41}$$

and the thermal conductivity

$$\chi = K/\rho_0 C_p. \tag{3.42}$$

The energy equation (3.40) makes explicit the direct coupling between the thermal fluctuations and the velocity field.

The pressure has no thermodynamic significance: this results from the assumption $d_3 \ll H$ since, from Eqns. 3.10 and 3.12,

$$\hat{p}'/p_0 \sim (d_3/H)(\hat{T}'/T_0) \ll \hat{T}'/T_0. \tag{3.43}$$

A pressure force is needed to drive the horizontal flow but it is determined solely by the continuity equation. Cross-differentiating Eqn. 3.34 eliminates the pressure. The motion can be solved in terms of the velocity and temperature fields $\mathbf{v}'$, T' alone and the pressure can be found subsequently by substitution in the original form (3.34) which has become a pressure balance equation. The mathematical advantages of this case are readily apparent but are gained at the cost of a drastic physical simplification.

Having thus constructed simplified mathematical models suitable for the description of motion driven by thermal forces, we can now examine in more detail the corresponding physical processes occurring. We look first at the onset of convection in various systems. This problem was first analysed by Lord Rayleigh (1916) and is therefore associated with his name. Rayleigh's purpose was to explain the results of a laboratory experiment so his classical analysis refers to a Boussinesq fluid heated from below.

This is not the only means by which convection may be initiated. A dense fluid overlying a less dense fluid forms an unstable system. The dense fluid sinks and the less dense rises – the so-called Rayleigh–Taylor instability – and these motions can also carry heat (see Section 3.4.5). Another unstable system arises if a horizontal fluid layer is made to flow sufficiently quickly over a denser layer – the Kelvin-Helmholtz instability. In the following pages we shall encounter these two instabilities only in passing. We shall concentrate on the classical Rayleigh problem, where the temperature gradient, rather than the density or velocity gradient, is primarily responsible for the onset of convection. At the same time we shall look at the extension of Rayleigh's analysis to stratified fluids.

3.3 The Rayleigh problem

3.3.1 Introduction

Let us consider the stability of a motionless fluid layer, heated from below, in a gravitational field. We suppose that the fluid is compressible and dissipative through the agencies of viscosity and heat conduction. In the following we shall explore the basic physical processes involved by means of a simple heuristic argument due to Normand, Pomeau & Valverde (1977).

Suppose that, at some instant, the fluid has a mean density stratification $\bar{\rho}(x_3)$ and that we move a parcel of fluid of radius R steadily upwards with speed v_3'. The velocity must be small in the sense already discussed; the surrounding medium must not be significantly disturbed and the vertical component of the pressure force must remain negligible.

Under stellar conditions the diffusive heat transport far outweighs the viscous heating. Eqn. 3.4*d* then yields a characteristic time for any temperature difference between the parcel and its surroundings to be smoothed out, this characteristic, or relaxation, time being $\sim R^2/\chi$. During this time the parcel will have travelled a distance $\Delta x_3 \sim v_3' R^2/\chi$ to a point where, in the absence of thermal exchange, its density would be

$$\rho(x_3 + \Delta x_3) = \bar{\rho}(x_3) + \overline{(\mathrm{d}\rho/\mathrm{d}x_3)_{\mathrm{ad}}}\,\Delta x_3, \tag{3.44}$$

$\overline{(\mathrm{d}\rho/\mathrm{d}x_3)_{\mathrm{ad}}}$ being the mean adiabatic density gradient. But the density of the surrounding medium is

$$\bar{\rho}(x_3 + \Delta x_3) = \bar{\rho}(x_3) + (\mathrm{d}\bar{\rho}/\mathrm{d}x_3)\,\Delta x_3, \tag{3.45}$$

so the parcel will experience a buoyancy force

$$\begin{aligned} -R^3 g\,[\rho(x_3 + \Delta x_3) - \bar{\rho}(x_3 + \Delta x_3)] \\ = R^3 g\,[\mathrm{d}\bar{\rho}/\mathrm{d}x_3 - \overline{(\mathrm{d}\rho/\mathrm{d}x_3)_{\mathrm{ad}}}]\,\Delta x_3 \\ = -R^3 \bar{\rho} N^2 (v_3' R^2/\chi), \end{aligned} \tag{3.46a}$$

with

$$N^2 = -(g/\bar{\rho})[\mathrm{d}\bar{\rho}/\mathrm{d}x_3 - \overline{(\mathrm{d}\rho/\mathrm{d}x_3)_{\mathrm{ad}}}]. \tag{3.47a}$$

This overestimates the buoyancy since energy exchange occurs during the upward motion of the parcel; this reduces the temperature and hence density difference between the parcel and its surroundings. Therefore the buoyancy force is in fact

$$-R^3\bar{\rho}N^2(v_3' R^2/\chi)\epsilon, \tag{3.46b}$$

where $0 < \epsilon < 1$.

When the mean density stratification is subadiabatic, the displaced parcel is denser than its surroundings and the negative 'buoyancy' becomes a restoring force. This causes the parcel to oscillate at a natural frequency N - the Brunt-Väisälä frequency.

When the mean density stratification is superadiabatic, N becomes imaginary so that, in an inviscid fluid, the displacement continues to grow. The sign of N^2 thus provides a criterion for convective instability, named after its discoverer K. Schwarzschild (1906). In real fluids, however, the buoyancy force is resisted by viscous drag. This can be estimated from Eqn. 3.2*b*, ignoring the pressure terms, as $R^3(\nu\bar{\rho}v_3'/R^2)$. The parcel will continue to accelerate only if the buoyancy outweighs the drag, i.e. if

$$-\bar{\rho}(N^2 v_3'/\chi)R^5\epsilon > \bar{\rho}\nu v_3' R. \tag{3.48}$$

The ratio of buoyancy to drag increases rapidly with parcel size; thus the greatest imbalance between buoyancy and drag occurs for the largest scale possible in the system, namely the thickness of the fluid layer d_3. If the superadiabatic gradient is slowly increased, this scale is the first to become unstable; this happens when

$$-N^2 d_3^4/\chi\nu \sim 1/\epsilon > 1. \tag{3.49}$$

This criterion can be written in its more usual form by transforming from density gradient to temperature gradient using the thermodynamic relation derived from the equation of state

$$\mathrm{d}\bar{\rho}/\mathrm{d}x_3 = (\bar{\rho}\alpha/\bar{p})\,\mathrm{d}\bar{p}/\mathrm{d}x_3 - (\bar{\rho}\delta/\bar{T})\,\mathrm{d}\bar{T}/\mathrm{d}x_3, \tag{3.50}$$

and recalling that there is presumed to be no pressure difference between the parcel and its surroundings. The Brunt-Väisälä frequency is now

$$N^2 \sim [\mathrm{d}\bar{T}/\mathrm{d}x_3 - \overline{(\mathrm{d}T/\mathrm{d}x_3)}_{\mathrm{ad}}]g\delta/\bar{T} = -g\delta\beta/\bar{T}, \tag{3.47b}$$

and the condition for instability takes the usual form of the Rayleigh criterion

$$\mathscr{R}_a = (\delta/\bar{T})(g\beta d_3^4/\chi\nu) = (\delta/\bar{T})g\beta d_3^4(\bar{\rho}C_p/K)(\bar{\rho}/\eta) > 1/\epsilon. \tag{3.51}$$

The left-hand side is a dimensionless number, the Rayleigh number, which measures the ratio of the buoyancy force to the viscous drag and which alone determines the stability of the fluid to convective motions. The condition can be expressed alternatively in terms of energy. Convection sets in when the rate at which free energy liberated by the rising hot elements is greater than the rate at which energy is dissipated by thermal conduction and viscous damping.

The Schwarzschild criterion for the onset of convection, to which the Rayleigh criterion reduces when the viscosity is vanishingly small, can be written as either

$$N^2 < 0 \tag{3.52a}$$

or

$$-\mathrm{d}\bar{T}/\mathrm{d}x_3 > -\overline{(\mathrm{d}T/\mathrm{d}x_3)_{\mathrm{ad}}}. \tag{3.52b}$$

Another form can be produced by appealing to hydrostatic equilibrium in the mean atmosphere. This allows us to eliminate the spatial gradient by virtue of

$$\mathrm{d}\bar{T}/\mathrm{d}x_3 = (\mathrm{d}\bar{T}/\mathrm{d}\bar{p})(\mathrm{d}\bar{p}/\mathrm{d}x_3) = -g\bar{\rho}(\mathrm{d}\bar{T}/\mathrm{d}\bar{p}), \tag{3.53}$$

or

$$\mathrm{d}\bar{T}/\mathrm{d}x_3 = -(g\bar{\rho}\bar{p}/\bar{T})\,\nabla,$$

where $\nabla = \mathrm{d}(\ln T)/\mathrm{d}(\ln p)$ is the logarithmic gradient. The Schwarzschild criterion can then be rewritten

$$\nabla > \nabla_{\mathrm{ad}}. \tag{3.52c}$$

In one form or the other, this inequality is commonly used in calculating solar and stellar models (Schwarzschild, 1958; Cox & Guili, 1968). Strictly, it is a necessary but not a sufficient condition. In stellar envelopes it is adequate. There the magnitudes of the other factors, particularly the vanishingly small atomic viscosity, ensure that an almost infinitesimal superadiabatic gradient will fulfil the Rayleigh criterion for the onset of convection; so the two criteria are effectively the same.

3.3.2 *Evaluation of the Rayleigh number*

The evaluation of the Rayleigh number for the *onset* of convection, the critical Rayleigh number $\mathscr{R}a_{\mathrm{c}}$, requires a fuller specification of the problem. Lord Rayleigh considered a bounded layer of Boussinesq fluid with a constant temperature difference maintained between its upper and lower boundaries. These circumstances allow the equations to be rewritten with scaled variables in a manner that throws a clear light on the interplay of the various terms.

Using appropriate characteristic dimensions (cf. Chandrasekhar, 1961) Eqns. 3.33-5 assume the dimensionless form

$$\partial\tilde{v}_k'/\partial\tilde{x}_k = 0 \tag{3.54}$$

$$\partial\tilde{v}_i'/\partial\tilde{t} + \tilde{v}_k'\,\partial\tilde{v}_i'/\partial\tilde{x}_k = (\nu/\chi)(-\partial\tilde{p}'/\partial\tilde{x}_i + \tilde{T}'\delta_{i3} + \partial^2\tilde{v}_i'/\partial\tilde{x}_k^2) \tag{3.55}$$

$$\partial\tilde{T}'/\partial\tilde{t} + \tilde{v}_k'\,\partial\tilde{T}'/\partial\tilde{x}_k = \mathscr{R}a\;\tilde{v}_3' + \partial^2\tilde{T}'/\partial\tilde{x}_k^2. \tag{3.56}$$

As noted before, cross-differentiating Eqn. 3.55 eliminates the scaled pressure $\tilde{p}'$

from the Boussinesq equation and cross-differentiating again produces the convenient form

$$\frac{\partial}{\partial \tilde{t}}\left(\frac{\partial^2 \tilde{v}_i'}{\partial \tilde{x}_k^2}\right) - \frac{\partial^2}{\partial \tilde{x}_i\,\partial \tilde{x}_j}\left(\tilde{v}_k'\frac{\partial \tilde{v}_j'}{\partial \tilde{x}_k}\right) + \frac{\partial^2}{\partial \tilde{x}_k^2}\left(\tilde{v}_j'\frac{\partial \tilde{v}_i'}{\partial \tilde{x}_j'}\right)$$

$$= \frac{\nu}{\chi}\left(\delta_{i3}\frac{\partial^2 \tilde{T}'}{\partial \tilde{x}_k^2} - \frac{\partial^2 \tilde{T}'}{\partial \tilde{x}_i\,\partial \tilde{x}_3} + \frac{\partial^4 \tilde{v}_i'}{\partial \tilde{x}_k^4}\right). \tag{3.57}$$

The fluid motion is thus characterized by just two parameters, the Rayleigh number based on the simple temperature gradient β,

$$\mathcal{R}a = (\delta/T)(g\beta d_3^4/\nu\chi), \tag{3.58}$$

and the Prandtl number

$$\mathcal{P}r = \nu/\chi. \tag{3.59}$$

The Prandtl number depends only on the fluid properties and measures the contribution of the inertial forces to the local acceleration relative to all the other forces. Its full significance will become apparent in Section 3.4.3.

For an ideal gas $\delta = 1$ so the Rayleigh number would vary with depth, but in most laboratory fluids $\delta \propto T$, and $\mathcal{R}a$ is constant. We treat the latter case here.

Just beyond the critical Rayleigh number the fluctuating quantities are small, and the non-linear terms involving them can be neglected. We need solve only the linearized equations. These (and only these) admit independent modal solutions with periodic horizontal planforms and exponential time dependence:

$$X(\tilde{\mathbf{x}}, \tilde{t}) = X(\tilde{x}_3)\exp(\tilde{n}\tilde{t})\exp[\mathrm{i}(\tilde{k}_1\tilde{x}_1 + \tilde{k}_2\tilde{x}_2)]. \tag{3.60}$$

Here it may be noted that, since such solutions generally form a complete set, a series expansion of linear modes can be employed when we seek to solve the general non-linear equations. However, the non-linear terms then couple the various modes together so that no particular mode can be isolated and determined independently of the others. Modal analysis (see Section 4.3.2) examines the truncation of such systems and can be employed when the state of the fluid is far beyond the onset of convection.

From the linearized equations, 3.56 and 3.57, one can eliminate $\tilde{T}'$ to produce a single sixth-order differential equation for the vertical velocity perturbation $\tilde{v}_3'$,

$$(\tilde{\nabla}_3^2 - \tilde{k}_\perp^2)(\tilde{\nabla}_3^2 - \tilde{k}_\perp^2 - \tilde{n})(\tilde{\nabla}_3^2 - \tilde{k}_\perp^2 - \tilde{n}\,\mathcal{P}r)\tilde{v}_3' = -\,\mathcal{R}a\,\tilde{k}_\perp^2\tilde{v}_3', \tag{3.61}$$

where the operator $\tilde{\nabla}_3 = \mathrm{d}/\mathrm{d}\tilde{x}_3$ and

$$\tilde{k}_\perp^2 = \tilde{k}_1^2 + \tilde{k}_2^2, \tag{3.62}$$

$\tilde{k}_\perp$ being the scaled horizontal wavenumber.

The fluid is unstable if any mode has a positive growth rate, i.e. if the growth rate $\tilde{n}$ is real and positive. As $\tilde{n}$ tends to zero, the fluid becomes marginally unstable, a state in which the disturbance neither grows nor decays in time. To find this transition state we set $\tilde{n}=0$, and look for solutions of the homogeneous equation

$$(\tilde{\nabla}_3^2-\tilde{k}_\perp^2)^3\tilde{v}_3' = -\mathcal{R}a\,\tilde{k}_\perp^2\tilde{v}_3'. \tag{3.63}$$

The Rayleigh number corresponding to the state of marginal stability appears as an eigenvalue in this problem. The Prandtl number does not appear and does not influence the *onset* of convection.

There is a series of solutions of Eqn. 3.63 for each choice of horizontal wavenumber $\tilde{k}_\perp$. We have already noticed that convection appears first at the largest scale possible, so there is a *minimum* eigenvalue, or Rayleigh number, for which the motion produces a single cell extending from one boundary to the other. (The vertical velocity eigenfunction vanishes only at the upper and lower boundaries.) This is the fundamental mode. Higher eigenvalues produce overtones in which the motion breaks up into a series of cells overlying one another. The dependence of the fundamental Rayleigh number on horizontal wavenumber is shown in Fig. 3.1. It has a well-defined minimum, termed the critical Rayleigh number $\mathcal{R}a_c$. As the temperature difference between the horizontal boundaries of an infinitely extended fluid layer is increased from zero, convec-

Fig. 3.1. Stability diagram for a low Prandtl number fluid. Below the curve the fluid is convectively stable, motions first setting in at the critical point $\mathcal{R}a_c$, $\tilde{k}_{\perp c}$. In region II, stable two-dimensional rolls form. The rolls are disrupted by further rolls, at right angles to the original in region III, and at slight angles in region IV (after Busse, 1967).

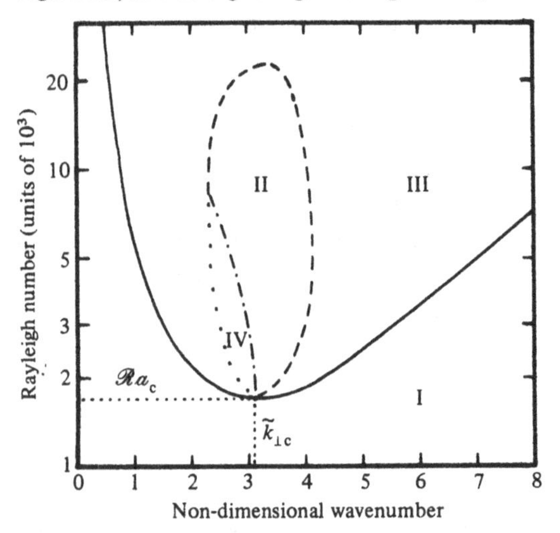

tive instability will first set in when $\mathcal{R}a_c$ is reached and the horizontal scale of the resulting motions will be $\Lambda_{\perp c} = 2\pi/k_{\perp c}$.

The evaluation of these quantities is a comparatively simple exercise and lends itself to almost endless variation. Table 3.2 compares the critical Rayleigh numbers and wavenumbers for various types of boundary conditions, taken from Chandrasekhar (1961).

Most laboratory experiments have rigid boundaries where viscous forces ensure that the motion vanishes, i.e. at the boundaries

$$v_1' = v_2' = v_3' = 0. \tag{3.64}$$

In the neighbourhood of the walls a viscous boundary layer develops which dominates the flow. This is absent under most astrophysical circumstances where free boundary conditions obtain. These require the vertical velocity component and the surface stresses to vanish, i.e.

$$v_3' = \partial v_1'/\partial x_2 = \partial v_2'/\partial x_1 = 0 \tag{3.65}$$

at the boundaries.

A more complicated Boussinesq problem with some features of astrophysical relevance was analysed by Stix (1970). He considered the unstable layer to be overlain by a stable semi-infinite layer of another Boussinesq fluid. If this second layer is isothermal, enforced by specifying infinite conductivity, the unstable layer is further destabilized. The critical Rayleigh number is reduced to 225 and the critical wavenumber to 1.25. On the other hand, adding a layer with even more stable stratification stabilizes the system in the sense that the critical Rayleigh number is increased to 2434.6 and the critical wavenumber to 3.63. In both cases the convective motion, after onset, extends beyond the unstable layer into the overlying fluid. This is an important feature of the boundary between stable and unstable fluid layers, which persists in finite-amplitude convection such as that occurring in the Sun (see Section 3.4.5).

An exhaustive discussion of the onset of convection in Boussinesq fluids, including the influences of rotation and magnetic fields, can be found in Chandrasekhar's (1961) monograph.

Table 3.2. *Critical Rayleigh numbers and wavenumbers for the onset of convection in Boussinesq fluids with various boundary conditions*

Boundary	$\mathcal{R}a_c$	$\tilde{k}_{\perp c}$
Both free	657.5	2.22
One free, one rigid	1100.6	2.68
Both rigid	1707.8	3.12

3.3.3 Extension to stratified fluids

We turn now to stratified fluids, where the literature is considerably less exhaustive. The onset of convection in polytropes lends itself to an analysis similar to that above, but employing the anelastic equations (Unno, Kato & Makita, 1960). Since the state of marginal stability is time-independent, its description in the anelastic approximation is identical to that in the general hydrodynamic equations. Spiegel (1965) derived a dimensionless equation, equivalent to Eqn. 3.63, describing the marginal stability of a perfect gas polytrope confined between boundaries at $-x_{03}$ and $-(x_{03}+d_3)$. In the same notation as before, it takes the form

$$\tilde{x}_3^{-m}(\tilde{\nabla}_3^2-\tilde{k}_\perp^2)\tilde{x}_3^{-m}\mathscr{L}\tilde{v}_3' = -\,\mathscr{R}a(\tilde{x}_{03})(\tilde{x}_{03})^{1-2m}\tilde{k}_\perp^2\tilde{v}_3', \tag{3.66}$$

where the differential operator $\mathscr{L}$ is

$$\tilde{x}_3(\tilde{\nabla}_3^2-\tilde{k}_\perp^2)^2+\tilde{x}_3(\tilde{\nabla}_3^2-\tilde{k}_\perp^2)\tilde{\nabla}_3(m/\tilde{x}_3)-(m+1)(\tilde{\nabla}_3^2-\tilde{k}_\perp^2)(\tilde{\nabla}_3+m/\tilde{x}_3)$$
$$+\tilde{k}_\perp^2 m(m+1)/(3\tilde{x}_3). \tag{3.67}$$

This is again a sixth-order homogeneous ordinary differential equation. The additional complexity introduced by even the simplest stratification is immediately apparent.

In fluids with spatially varying properties no unique Rayleigh number can be defined. In a polytrope the Rayleigh number based on the total layer thickness varies with height as

$$\mathscr{R}a\,(\tilde{x}_3)=(1/T_0)(g\beta d_3^4)(\rho_0 C_p/K)(\rho_0/\eta)=\mathscr{R}a\,(\tilde{x}_{03})(-\tilde{x}_3/\tilde{x}_{03})^{2m-1}, \tag{3.68}$$

if η and K are taken to be constant throughout the perfect gas ($\delta = 1$). The value $\mathscr{R}a_c(\tilde{x}_{03})$ can be found by solving the eigenvalue problem, whence the critical Rayleigh number at each depth can be evaluated. Gough, Moore, Spiegel & Weiss (1976) took $\mathscr{R}a$ at the midpoint of the polytropic layer to be representative. Their results for the case of free boundaries are summarized in Table 3.3, which gives the critical Rayleigh numbers and wavenumbers at the

Table 3.3. *Critical Rayleigh numbers and wavenumbers for the onset of convection in polytropes of varying polytropic index (m) and thickness ($\tilde{x}_{03}$)*

$\tilde{x}_{03}$	10.0		1.0		0.1		0	
m	$\mathscr{R}a_c$	$\tilde{k}_{\perp c}$	$\mathscr{R}a_c$	$\tilde{k}_{\perp c}$	$\mathscr{R}a_c$	$\tilde{k}_{\perp c}$	$\mathscr{R}a_c$	$\tilde{k}_{\perp c}$
0.5	658	2.22	690	2.24	930	2.35	1144	2.42
1.5	660	2.22	762	2.26	1381	2.53	1671	2.69
2.5	661	2.22	837	2.30	1426	2.87	1481	3.13
3.5	663	2.22	902	2.37	1118	3.33	974	3.68

midpoints of layers with various polytropic indices and thicknesses. Note that in dimensionless units the boundaries are at $-\tilde{x}_{03}$ and $-(\tilde{x}_{03}+1)$, so varying thicknesses are achieved by changing $\tilde{x}_{03}$.

Large values of the polytropic index result in layers with a large density variation between the boundaries (strong stratification). These tend to be generally more stable than layers of the same thickness with slight density variation. Strong stratification implies relatively low densities at the top of the layer with concomitant large kinematic viscosity and thermal conductivity, which provide the stabilization. At the onset of convection the motions in the upper layers suffer larger damping than those in the lower, counteracting the tendency of momentum conservation to generate high speeds there (see Sections 4.3.2 and 4.3.3).

Solutions for $\tilde{x}_{03}=0$, i.e. in which the outer boundary coincides with the surface of the polytrope where the pressure and density vanish, are particularly relevant to free atmospheric convection; in this case, strong stratification is also stabilizing. The *apparent* destabilization indicated in Table 3.3 is simply the result of choosing to evaluate $\mathscr{R}a$ at the midpoint of the layer. But it is noteworthy that the range of critical values for the onset of convection in different polytropes is no greater than the range of values for Boussinesq fluids with different types of boundary. The horizontal wavelength $\Lambda_{\perp c}$ of the first unstable mode is always approximately

$$\Lambda_{\perp c}=2\pi d_3/\tilde{k}_{\perp c}\sim 3d_3. \tag{3.69}$$

Thus the horizontal size of the initial structure that develops in a convective layer is *comparable to the vertical extent of the layer whether that layer is stratified or not*.

More general discussions of the onset of convection in compressible (non-polytropic) atmospheres may be found in Vickers (1971) (but see also Gough *et al.* (1976)) and Graham & Moore (1978). Since we shall see in Section 3.4.4 that the convective state in stars is far removed from that of marginal stability, such detailed analyses need not detain us.

The linearized analysis described above has been put to a further use which warrants brief comment. If we could conceive of a fluid being released from rest with a Rayleigh number greater than the critical value (a difficult experiment!), the linearized equations could be employed to follow the initial motion. Their solution yields the growth rate $\tilde{n}$ of each mode as a function of its horizontal wavenumber $\tilde{k}_{\perp}$. At least one mode would be growing since the system, by assumption, is beyond the point of instability; growing modes can be guaranteed by neglecting the viscosity ($\mathscr{R}a\sim\infty$). Such modes have been investigated in detail for polytropic atmospheres by Skumanich (1955), Böhm & Richter (1959,

1960) and Spiegel (1964). But it is important to realize that these cellular solutions are *not* stationary flows. They describe only the initial motion when the amplitude grows exponentially before non-linear limiting effects become apparent. There are no grounds for supposing that the mode with the largest initial growth rate will dominate the motion in stationary finite-amplitude convection. Several studies have sought, explicitly or implicitly, to prove this supposition, but without success. Linearized solutions tell us nothing about finite-amplitude convection, such as that responsible for the solar granulation.

3.4 Laboratory convection

3.4.1 Introduction

Rayleigh's theoretical investigation of the onset of convection was inspired by the laboratory experiments of H. Bénard which were carried out at the beginning of the present century. Bénard worked with thin liquid layers, about 1 mm deep, standing on a heated metal plate kept at a uniform temperature. The upper surface was usually a free boundary in contact with the air. Bénard employed a number of liquids in his experiments; some of these are solids at ordinary temperatures. The outstanding result of Bénard's experiments was that the fluid layer resolved into a cellular pattern if the temperature gradient was sufficiently large. This pattern consisted of somewhat irregular polygons in the initial stages, but when the heating was prolonged became a system of regular hexagons. Not only did this work stimulate Rayleigh but in addition it led H. H. Plaskett in 1936 to identify the granulation with cellular convection.

It is ironic that these important developments were based on a fundamental misinterpretation of Bénard's results. More recent experimental (Block, 1956) and theoretical (Pearson, 1958) work has clearly demonstrated that Bénard's convection was not driven by buoyancy but by surface tension forces, which Bénard himself suspected. Striking confirmation comes from the experiments performed aboard the Apollo 14 and 17 spacecraft in which gravity, and hence buoyancy, was reduced to only 10^{-6}–10^{-8} of its value in terrestrial experiments. Yet the typical Bénard cells showed little change (Grodzka & Bannister, 1975).

We must turn to other experiments for confirmation of Rayleigh's predictions for the onset of convection. The results most relevant for comparison with observations of the solar granulation are, of course, those referring to true buoyancy-driven convection. There is a wealth of literature on such experiments but almost all of it deals with incompressible fluids, usually strictly Boussinesq ones. To avoid surface tension forces the fluid must be bounded both above and below; experimental convenience dictates the use of rigid plates to contain the convecting fluid, although Goldstein & Graham (1969) describe an experiment performed with free boundaries.

Experimental apparatus always has side walls. In most experiments the horizontal extent is made greater than the depth of the fluid, an arrangement said to have a high aspect ratio. This to some extent mimics the unbounded situation in free convection but the degree to which the results are influenced by the side walls is uncertain. Although such conditions differ widely from the astrophysical regime, no other system has been studied so fully.

For detailed reviews of convection in the laboratory the reader is referred to Whitehead (1971), Koschmieder (1974) and Palm (1975).

3.4.2 The onset of convection

As we have seen, the Rayleigh analysis predicts that fluids subject to a (very slowly) increasing vertical temperature gradient will remain motionless until a critical Rayleigh number is reached, whereupon convective motions set in with amplitudes that grow rapidly with time. In experiments at Rayleigh numbers well below the critical value predicted by linear analysis the fluid does indeed remain static; but as that number is approached convective motions develop gradually, instead of suddenly. It would appear that a small, but unavoidable, heat transfer through the side walls drives a circulation which in turn triggers finite-amplitude convective instabilities below the critical Rayleigh number of linear theory (Daniels, 1977; Hall & Walton, 1977, 1979). However, it is doubtful whether such effects would occur in natural unbounded fluids. Apart from such smoothing of the onset, the behaviour of convection just above the critical Rayleigh number is well described by linear theory.

In systems maintained with a slightly supercritical Rayleigh number the convective motion appears first in the form of horizontal two-dimensional rolls which gradually evolve until they acquire a steady character. Not surprisingly, the final orientation of the rolls is determined by the geometry of the container. For most laboratory fluids these rolls are the only stable marginally convective configuration. This has been proved theoretically for Boussinesq fluids by Schlüter, Lortz & Busse (1965). The stability of the rolls is easy to understand: there is complete symmetry between the upward and downward motions, which reflects the uniformity of the fluid properties. In fluids with spatially varying properties symmetry between up and down cannot be expected and cellular flows, in which the topology of the upward and downward motions is different, may be stable.

For example, Palm (1960) and Segel & Stuart (1962) have demonstrated that, if the variation of viscosity with temperature is sufficiently large, hexagonal cells may be a stable configuration. The direction of flow at the centre of the cell is upwards or downwards depending on whether the viscosity decreases or increases with temperature (Palm, 1960; Stuart, 1964). Liquids generally possess a viscosity that decreases with increasing temperature, so their convection cells have upflow at the centre and downflow at the edges; in most gases, the situation is reversed. The flow pattern in the solar gas, however, is the same as in a liquid and the explanation for the observed direction of flow must be sought elsewhere (see Section 5.3).

By increasing the temperature difference between the lower and upper boundaries beyond the critical value necessary to generate motion, we can produce convection at any Rayleigh number we choose. In most laboratory fluids the convective rolls at first simply increase in width as the Rayleigh number is increased beyond the critical value, but at $\mathscr{Ra} \sim 12\,\mathscr{Ra}_{\mathrm{c}}$ most fluids develop another dynamic instability which leads to steady three-dimensional flow. This takes the form of bi-modal orthogonal rolls which give rise to a rectangular cellular pattern (Krishnamurti, 1970a, 1973).

The change from two- to three-dimensional flow can also be seen in the behaviour of the heat flux, which is one of the most easily measured basic parameters in convection experiments. It can be used to define another dimensionless number, the Nusselt number, $\mathscr{Nu}$, which is the ratio of the total measured flux to the purely *conductive* flux which would occur for a constant temperature gradient between the two boundaries. It thus measures the 'degree' of convection. Whilst a particular pattern of motion persists, the degree of convection is found to increase smoothly with the Rayleigh number: but each change of pattern coincides with a change of slope. These heat-flux transitions can be seen by plotting the heat flux in the form of the product of the Nusselt and Rayleigh numbers (the Rayleigh number, being proportional to the temperature gradient, measures the conductive flux) as a function of the Rayleigh number. Malkus (1954) found 7 linear segments with steadily increasing slope for Rayleigh numbers in the range $1700 < \mathscr{Ra} < 10^6$. Willis & Dearsdorff (1967) have given a more complete account of the transitions, up to $\mathscr{Ra} = 2.8 \times 10^6$. Earlier, Malkus & Veronis (1958) attributed each transition to the onset of a new type of instability. This seems well established for the first few transitions, but general confirmation is difficult to obtain as the onset of the finite-amplitude instabilities appears to depend on the particular experimental arrangement and the way in which the experiment is conducted (Krishnamurti, 1970a).

Fig. 3.2 shows the heat flux plotted against the Rayleigh number for an oil (Krishnamurti, 1970a,b). The generally linear dependence and the abrupt changes of slope are well demonstrated; at the transition $\mathscr{Ra}_{\mathrm{II}}$ the convective pattern changes from two-dimensional rolls to three-dimensional cells.

Malkus & Veronis also proved that, for a particular class of system, the flow pattern would change to produce a more efficient heat transfer, implying that the slope of the heat flux would increase at each transition. This is certainly seen at low Rayleigh numbers, but at higher values the situation is not so clear. For instance, Krishnamurti (1973) finds decreases in slope for air and mercury at large $\mathscr{R}a$, which would imply a decrease in the efficiency of heat transport. The heat flux curve for mercury is shown in Fig. 3.3; the less clear-cut nature of the transitions is apparent.

However plausible the picture of successive instabilities, we note that the very existence of heat-flux transitions has been questioned by Koschmieder & Pallas (1974), who find only smooth curves. The subject of low Rayleigh number convection is by no means closed, so we should be wary as we pursue the development to higher values.

3.4.3 *The transition to developed convection*

When the convective flow is established a circulation develops in which the hot rising fluid is replenished by the cool sinking material. The fluid turns over, and the implied rotation suggests that the flow pattern can be usefully visualized in terms of the distribution of local rotation or angular momentum.

Fig. 3.2. Heat flux for an oil ($\mathscr{P}r = 100$) plotted against Rayleigh number. The heat flux is measured by the product of the Nusselt and Rayleigh numbers (see text). The transitions at which the nature of the flow is thought to change abruptly are marked (Krishnamurti, 1970a, b).

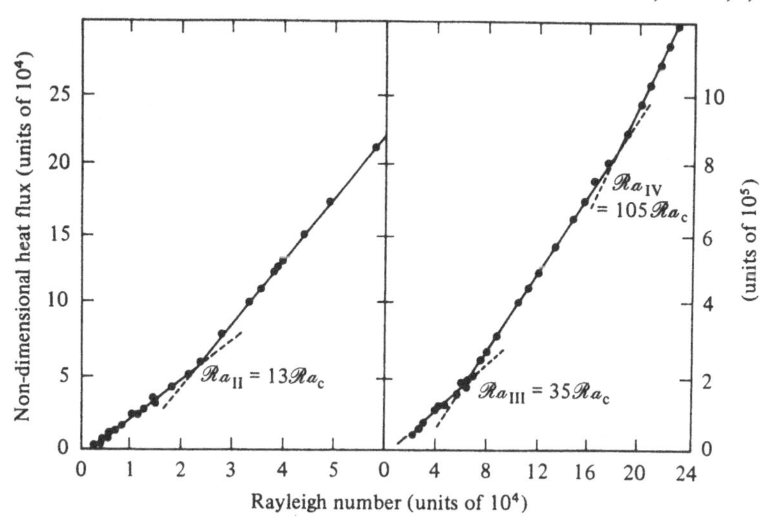

This is indeed a powerful tool and is described by the property of the fluid known as the vorticity.

The vorticity is defined as the curl of the velocity

$$\boldsymbol{\omega}' = \nabla \times \mathbf{v}', \tag{3.70}$$

and its significance is evident if we substitute this quantity in the inertial and viscous terms of Eqn. 3.39, giving

$$\frac{\partial \mathbf{v}'}{\partial t} = -\delta \mathbf{g}\frac{T'}{T_0} - \frac{1}{\rho_0}\nabla p' - \nabla(\tfrac{1}{2}v'^2) - \mathbf{v}' \times \boldsymbol{\omega}' - \nu\nabla \times \boldsymbol{\omega}. \tag{3.71}$$

Fig. 3.3. Heat-flux curve for mercury ($\mathscr{P}_r$ = 0.025). The heat flux is measured by the product of the Nusselt and Rayleigh numbers. Crosses refer to time-dependent flows (Krishnamurti, 1973).

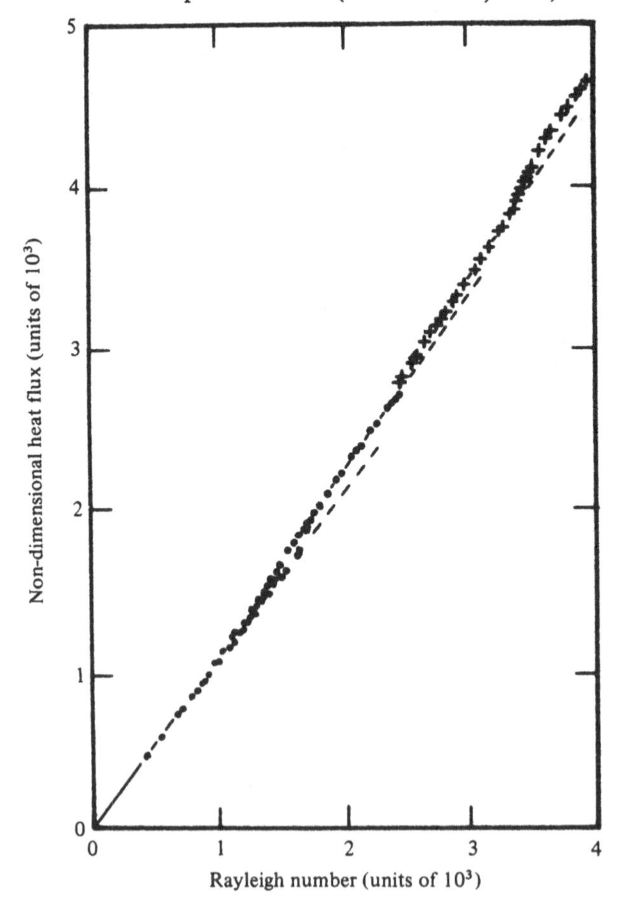

The vector product $\mathbf{v}' \times \boldsymbol{\omega}'$ – expressing the interaction of the flow with the vorticity – is equal to the Coriolis term that would be present if the coordinate system were rotating locally with angular velocity $2\boldsymbol{\omega}'$.

An 'equation of motion' for the vorticity can be found by cross-differentiating Eqn. 3.71 to yield

$$\partial \boldsymbol{\omega}'/\partial t = \nabla \times (\mathbf{v}' \times \boldsymbol{\omega}') + \nu \nabla^2 \boldsymbol{\omega}' + \nabla \times (-\delta T' \mathbf{g}/T_0 - \nabla p'/\rho_0). \qquad (3.72)$$

In a uniform, incompressible fluid the sources – the last two terms on the right-hand side – vanish and it is possible to maintain an irrotational flow in which the vorticity is always zero. In real fluids this is rarely the case.

The significance of Eqn. 3.72 is best seen by first ignoring the source terms. Then we have

$$\partial \boldsymbol{\omega}'/\partial t = \nabla \times (\mathbf{v}' \times \boldsymbol{\omega}') + \nu \nabla^2 \boldsymbol{\omega}' \qquad (3.73a)$$

and

$$\nabla \cdot \boldsymbol{\omega}' = \nabla \cdot (\nabla \times \mathbf{v}') \equiv 0. \qquad (3.74)$$

These constitutive relations for the vorticity are, in this special case, the same as those for the magnetic field in hydromagnetics.

By analogy with magnetic field lines, the fluid can be imagined to consist of vortex lines about which slender fluid elements rotate. Vortex lines form closed loops which can be advected by the fluid and can diffuse through it. Rewriting Eqn. 3.73*a* as

$$\partial \boldsymbol{\omega}'/\partial t + (\mathbf{v}' \cdot \nabla)\boldsymbol{\omega}' = (\boldsymbol{\omega}' \cdot \nabla)\mathbf{v}' + \nu \nabla^2 \boldsymbol{\omega}', \qquad (3.73b)$$

we see that the vortex lines are in general deformed as they are swept around. The first term on the right-hand side has contributions both from the shears $\partial v_i'/\partial x_j$ $(i \neq j)$, which tilt the vortex lines, and from $\partial v_i/\partial x_i$ (no summation implied), which stretch or compress the lines. Vorticity is increased by the stretching of a vortex line; conservation of angular momentum requires that the size of the rotating element then becomes correspondingly smaller. Diffusion counters this process.

In the low Rayleigh number experiments described in Section 3.4.2, the vertically directed buoyancy forces can generate only horizontal vorticity. This results in the formation of two-dimensional rolls. In strictly two-dimensional flow the horizontal vortex lines cannot be distorted and the buoyancy forces cannot produce a swirling motion in the horizontal plane. It is not surprising that this is a stable configuration for mildly supercritical flows.

In reality, however, fluctuations in the flow always occur, taking the form of three-dimensional perturbations. Whether these grow or decay is governed by the rate at which vorticity is increased by the inertial terms, i.e. by distortion and stretching, relative to the rate at which it is reduced by diffusion. As we can see from the dimensionless equation (3.55), the relative sizes of the viscous and inertial terms are governed by the Prandtl number $\mathcal{P}r$.

It should be remarked that this parameter arises naturally in convection-driven flows from our choice of scaling parameters. In many laboratory experiments, on the other hand, a shear flow is produced by forcing a fluid through a grid or tube or along a surface. If the characteristic spacing, or distance from the surface, is d and the mean flow speed v, the natural measure of the ratio of the inertial to viscous forces is

$$(v^2/d)(d^2/\nu v) = vd/\nu = \mathcal{R}e, \tag{3.75}$$

a dimensionless parameter known as the Reynolds number. In convective systems there is no imposed scale.

The Prandtl number determines how readily a two-dimensional flow can develop three-dimensional characteristics and, less obviously, whether the flow remains steady or becomes time-dependent. Steady three-dimensional flow results when the vorticity develops only in the horizontal plane. It requires any vertical component to be damped by viscosity, i.e. a high Prandtl number. If the Prandtl number is too low, all components of the vorticity develop simultaneously and a more complicated three-dimensional motion with a swirling component results. All these cases have been extensively investigated, both experimentally and, somewhat less thoroughly, theoretically.

In fluids such as water and oils ($\mathcal{P}r \geqslant 5$), the flow does not become time-dependent until after steady three-dimensional cells have developed (see $\mathcal{R}a_{\mathrm{III}}$ in Fig. 3.2). Non-steady behaviour results from an instability of the boundary shear layer at the horizontal surfaces. Krishnamurti (1970b, 1973) noted that small volumes of fluid with high shear detach themselves from the boundary and are advected around the cell. These elongated blobs are sometimes described as thermal plumes. The passage of successive blobs past a point is seen as an oscillation of the fluid parameters having a period equal to the cell turnover time, which decreases with increasing $\mathcal{R}a$. At $\mathcal{R}a_{\mathrm{IV}}$ (see Fig. 3.2) the first harmonic seems to appear. Eventually, the motion resolves itself into a stationary spoke-like pattern. This, in turn, at higher $\mathcal{R}a$ gives way to a phase in which cells continually form and then dissolve (Busse & Whitehead, 1971, 1974).

In fluids of lower Prandtl number, air for instance, the transition to time-dependence occurs at much lower Rayleigh numbers, and the stage of steady three-dimensional flow appears to be suppressed. However, there seems to be little agreement among experimenters on the actual pattern of development.

For air ($\mathcal{P}r = 0.71$), Krishnamurti (1970b, 1973) finds a rapid transition from steady two-dimensional rolls to propagating lateral oscillations at $\mathcal{R}a \sim 3.3\,\mathcal{R}a_{\mathrm{c}}$ and further changes marked by heat-flux transitions at $6\,\mathcal{R}a_{\mathrm{c}}$

and 10 $\mathcal{R}a_c$. Willis & Dearsdorff (1970) observe not only oscillations but also other non-steady phenomena - evolving horizontal flows and cell formation and dissolution - which can set in either before or after the appearance of oscillations in the rolls.

Ahlers & Behringer (1978), working with liquid helium, whose Prandtl number varies between 0.8 and 4 depending on temperature, find the aspect ratio of the apparatus to play a significant role. If the horizontal and vertical dimensions are comparable (i.e. Γ = radius/height = 2), they observe the onset of oscillations at $\mathcal{R}a \sim 10\ \mathcal{R}a_c$ and aperiodic fluctuations (a broad band of low frequencies) at $\mathcal{R}a \sim 11\mathcal{R}a_c$. At intermediate aspect ratios ($\Gamma = 5$), no oscillations are seen but broad-band temporal variations set in at $\mathcal{R}a \sim 2\mathcal{R}a_c$. In wide, shallow systems ($\Gamma = 57$), the broad-band fluctuations develop almost immediately after the onset of convection, $\mathcal{R}a \sim 1.25\mathcal{R}a_c$. The authors provide no illustrations of these flows.

Similar results are reported for mercury ($\mathcal{P}r = 0.025$) by Krishnamurti (1973); the corresponding heat-flux curve is shown in Fig. 3.3. Oscillations are seen briefly but are almost immediately superseded by multiply periodic behaviour. In this experiment, the aspect ratio was 25. These findings suggest that steady two- and three-dimensional motions in low $\mathcal{P}r$ fluids depend on an organization imposed by the physical boundaries of the container. In the absence of side walls we might expect all low $\mathcal{P}r$ fluids to show only disorganized behaviour.

The influence of the rigid surfaces at top and bottom is difficult to investigate experimentally as it involves comparison with results for free surfaces which, however, bring unwanted surface tension effects into play. This forces us back onto theoretical and numerical modelling. In particular, there is a large literature on the theory of Boussinesq fluids at low Rayleigh numbers. As noted before, Schlüter *et al.* (1965) proved that two-dimensional rolls are the only stable marginally convective configuration (cf. Fig. 3.1). Later works have treated the instabilities at higher Rayleigh numbers in fluids of moderate (Busse & Clever, 1979) and high (Clever & Busse, 1974) Prandtl number. These authors assume the fluid to be contained within rigid boundaries.

For convection between two free boundaries the results of numerical modelling are somewhat more extensive than those obtained by means of theoretical analysis. Moore & Weiss (1973a) pursued two-dimensional roll computations in slab geometry as far as $10^3\ \mathcal{R}a_c$ for fluids with $10^{-2} < \mathcal{P}r < 10^2$. Jones, Moore & Weiss (1976) assumed cylindrical geometry, with essentially similar results; steady solutions were found for all values of $\mathcal{R}a$. Only in fluids with $\mathcal{P}r > 1$ did Moore & Weiss encounter a heat-flux transition. It separates a viscous regime at low $\mathcal{R}a$, in which the vorticity is dissipated near the sites of its

Fig. 3.4. Plume structure in two-dimensional Boussinesq convection ($\mathscr{R}a = 1000\ \mathscr{R}a_c$, $\mathscr{P}_r = 6.8$, $\tilde{k}_\perp = 2\pi$) (Moore & Weiss, 1973a).
(a) Vertical profiles of temperature at cell centre, $\tilde{x}_1 = \frac{1}{2}$ (full line), and of temperature and vertical velocity at $\tilde{x}_1 = 1$ (dot-dashed and dashed lines respectively). The deviations from a constant temperature at cell centre are caused by the plumes at $\tilde{x}_1 = 0$ and 1 spreading out horizontally as they near the top and bottom boundaries.
(b) Horizontal profiles of temperature (full line) and vertical velocity (dashed line) at the mid-height of the cell ($\tilde{x}_3 = -0.5$). The cell rotates about its centre with almost uniform angular velocity (constant vorticity); the heat convection is concentrated in the plumes at $\tilde{x}_1 = 0$ and 1.

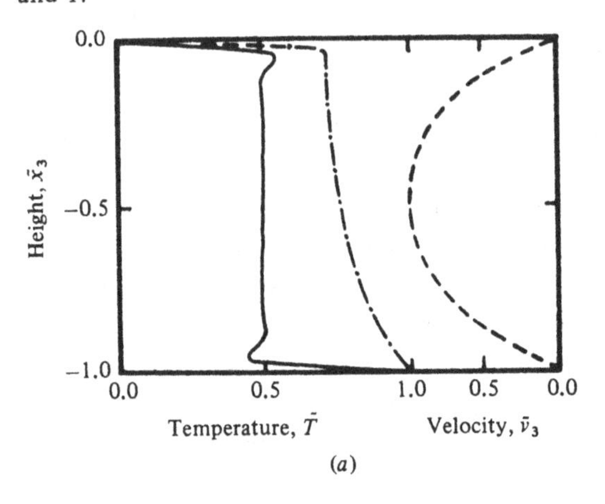

(a)

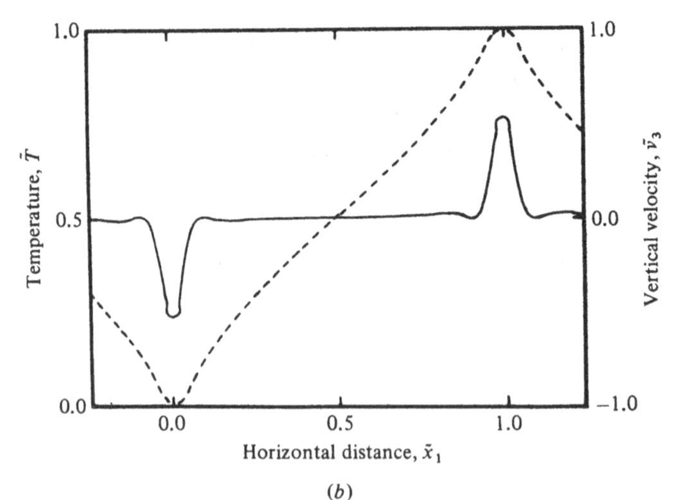

(b)

production, from an advective regime at higher $\mathscr{Ra}$, in which the vorticity is carried in thin plumes to the shear layers at the boundaries and dissipated preferentially near the corners of the cell. The temperature and vertical velocity profiles of these plumes are shown in Fig. 3.4. In water this transition should occur at $\mathscr{Ra} \sim 50\ \mathscr{Ra}_c$, when the rigidly bounded fluid would already have developed a three-dimensional flow pattern. But the boundary layers in the neighbourhood of free boundaries remain stable longer than those adjacent to rigid boundaries (Busse, 1967), so the two-dimensional solutions of Moore & Weiss are likely to remain stable to higher $\mathscr{Ra}$ than the experiments with rigid walls would indicate.

In fluids of lower $\mathscr{Pr}$ (<1), the viscosity is not sufficient to dissipate the vorticity in one cell turnover time. By confining ourselves to two dimensions, we deny the fluid any opportunity to transfer kinetic energy to small scales through vorticity changes. We thus force the velocities to increase until the large-scale shears throughout the fluid are sufficient for the viscous losses to balance the vorticity production. This, however, is physically unrealistic. The inertial effects in low $\mathscr{Pr}$ fluids cause the two-dimensional freely bounded flow to become unstable to three-dimensional disturbances close to $\mathscr{Ra}_c$ (Busse, 1972), just as in rigidly contained convection.

The instability of the numerical axisymmetric solutions was explored by Jones & Moore (1979). They found that a non-axisymmetric perturbation started to grow when the inertial terms became some 300 times greater than the viscous terms. This dynamical instability had no oscillatory component but indicated a direct fission of the cell, though a perturbation analysis cannot be followed through to this stage. For low Prandtl number ($\mathscr{Pr} = 0.01$), the most unstable mode would lead to three fragments in the range of Rayleigh numbers considered ($1.1\,\mathscr{Ra}_c < \mathscr{Ra} < 6\,\mathscr{Ra}_c$) for a cell whose aspect ratio (radius/depth) is 1.75. Wider cells show a tendency to split into more fragments. These results leave no doubt that low $\mathscr{Pr}$ fluids must be studied by three-dimensional modelling.

No numerical Boussinesq calculations for the three-dimensional, free boundary case appear to have been carried out. For rigid boundaries, exact results are limited to mildly supercritical convection, $\mathscr{Ra} < 15\,\mathscr{Ra}_c$ (see Weiss, 1977). Simplified models in which a sinusoidal horizontal variation of given wavelength is prescribed have been used to look for steady-state solutions up to very large $\mathscr{Ra}$ (Murphy, 1971; Gough, Spiegel & Toomre, 1975).

These calculations were designed to explore the dependence of the heat flux on the Rayleigh and Prandtl numbers. The asymptotic form for the Nusselt number in the limit of very small Prandtl number was found to be

$$\mathscr{Nu} \propto (\mathscr{Ra} \cdot \mathscr{Pr})^2, \tag{3.76}$$

for both free and rigid boundaries. However, the authors did not investigate the evolution of the convecting system and its successive instabilities, let alone its detailed velocity and thermal structure.

At the present time the more complete picture of laboratory convection is provided by studies of convection between rigid boundaries. A convenient summary of the results of these studies is given in Fig. 3.5, adapted from Krishnamurti (1973).

3.4.4 Turbulent convection

When the Rayleigh number is increased, a point is reached at which viscosity is unable to prevent the growth of vorticity and its interaction with the flow, regardless of the Prandtl number. Eventually the geometry of the flow becomes so topologically complicated that the flow cannot be traced in detail. The flow takes on a stochastic character that we describe as turbulent.

The time dependence of such flows shows no selected frequencies, just a broad-band spectrum. The onset of such behaviour is denoted by the curve labelled V in Fig. 3.5. The range of Rayleigh numbers over which non-turbulent motions occur narrows drastically at lower Prandtl numbers. If we extrapolate the V-curve to even lower values, the conclusion is very striking: for $\mathscr{P}_r < 10^{-2}$,

Fig. 3.5. Flow regimes in Boussinesq convection as a function of the Rayleigh and Prandtl numbers. The Prandtl numbers of the most commonly used laboratory fluids are marked along the upper scale (Krishnamurti, 1973).

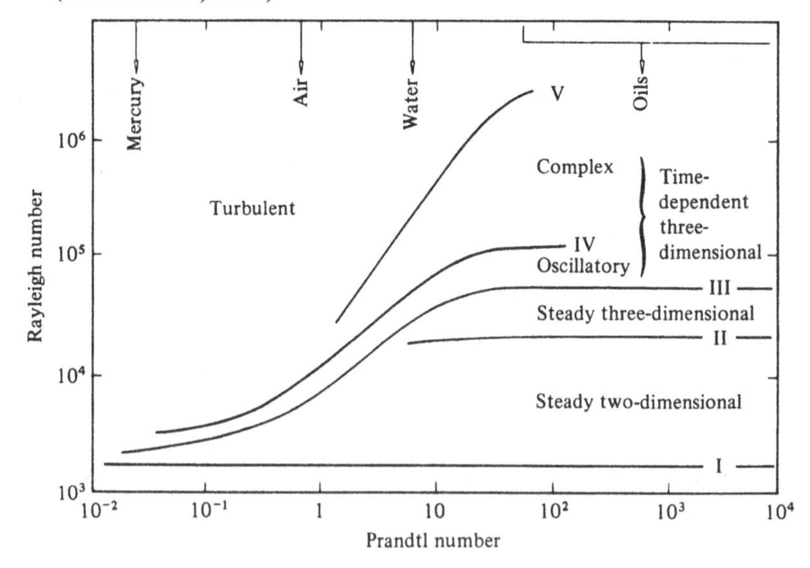

any convective flow in the laboratory becomes turbulent almost immediately. The discussion in Section 3.4.3 gives us no grounds to suspect that convection between free boundaries is any different.

Laboratory convection is comparatively simple, in that it depends on only two dimensionless numbers, $\mathscr{R}a$ and $\mathscr{P}r$. In the Sun, convection takes place in a compressible, stratified medium, and we must expect further parameters to be important. But the general physical reasoning which accounts for the special roles of $\mathscr{R}a$ and $\mathscr{P}r$ still applies, and these parameters must be critical factors determining the nature of solar convection. The evaluation of these parameters for the Sun is not an easy matter since we know that the accessible visible layers reflect only the very top of the convection zone. In these layers the local Prandtl and Rayleigh numbers can be determined from observation (Section 4.2.2) to be

$$\mathscr{P}r \sim 10^{-9},$$

$$\mathscr{R}a \sim 2 \times 10^{11}.$$

Evidently, therefore, we are dealing with a very low Prandtl number fluid and hence with turbulent convection.

Turbulent flows are exceedingly hard to analyse and to understand, but some general properties are quite readily grasped. Clear introductions to the subject are given by Tennekes & Lumley (1972) and Bradshaw (1976).

The *small-scale* motion of turbulent flows is fully stochastic. Any two points on a vortex line can be imagined to undergo a random walk. A property of random walks is the tendency of the end-points to drift apart, so that overall the vortex lines are stretched, increasing the vorticity and decreasing the rotational scale. This process continues until the viscous scales are reached, at which stage (vorticity) diffusion brings it to a halt. The viscous dissipation is roughly proportional to the mean square vorticity fluctuation, so the vorticity builds up until there is a rough balance between the kinetic energy production at large scales and viscous dissipation at small scales.

The kinetic energy transfer to smaller scales is known as a 'cascade'. A simple cascade occurs if there is a range of spatial scales over which there are no effective sources or sinks producing or destroying kinetic energy. Such scales are called 'inertial' scales since the energy transfer is due to the inertial term, and they are only indirectly coupled to the large-scale driving motions. The inertial scales carry no heat as, by definition, they have no energy sources and are not directed flows. On the contrary, their stochastic nature implies a tendency towards isotropy.

In an isotropic, homogeneous, inertial regime, simple dimensional arguments require the distribution of kinetic energy among the various spatial scales to follow the so-called Kolmogoroff law. The kinetic energy per unit spatial wave-number k is then

$$E(k) \propto k^{-5/3}. \tag{3.77}$$

The larger scales have the most energy and highest velocities, which must be the case for the energy to be fed down through the chain of smaller scales. Even in the absence of isotropy and homogeneity this steep decline will remain.

Like solar convection, most naturally occurring turbulent flows are far from being isotropic and homogeneous. The characteristic features of such flows - coherence, intermittency and entrainment - are of great importance but are little understood despite intensive investigation. A review of recent developments in the theory of turbulence will be found in Moffatt (1981).

Turbulence is never as homogeneous as theoreticians would like. There is always a tendency at the larger scales to form coherent structures (Liepmann, 1979). These are difficult to define, but the concept is intuitively clear: the flow shows patterns that are more spatially extended and longer-lived than one would expect as the result of a random superposition of motions of varying scale. Such organized flow regimes are particularly noticeable at the boundaries of turbulent regions.

Regions of turbulent flow are often limited to a well-defined volume or sheet - for example, the neighbourhood of a surface in shear flows. The vortex lines or sheets will then induce *irrotational* motion at other points of the fluid. But the boundary between the two types of flow is not hard and fast. Turbulent eddies come and go at the boundaries. The rotational or irrotational character of the fluid at a given location on the boundary is therefore intermittent.

In the irrotational region, the turbulence will decay by diffusion. The damping clearly increases with the smallness of the turbulent eddy, so large-scale eddies travel further. In the inverse process, the turbulent region will suck in and incorporate irrotational fluid. This exchange of material between regions of different turbulent intensity is called entrainment. It is an important feature of the dynamics of thermals in the Earth's atmosphere and of astrophysical models based on them (Section 4.2.3; Moore, 1967).

In astrophysics we deal with free convection, in which the motions are initiated simply by the imposed temperature gradient. The convection in the lowest levels of the Earth's atmosphere, described in Section 2.2.2, is forced. Winds, driven by the atmospheric pressure gradients, are strongly sheared at the Earth's surface. As in a pipe flow, a wind of sufficient strength will cause the flow in a boundary layer near the ground to become turbulent. The dynamically induced small-scale eddies also transport heat because the temperature of the ground is raised by solar irradiation, but they do not exist for that reason. Such boundary layers are not present in stars, so our discussion of turbulent convection in Chapter 4 will be devoted exclusively to the case of free convection.

3.4.5 *Overshoot in laboratory systems*

Suppose we have a region of stationary fluid confined between upper and lower boundaries maintained at constant temperatures. Furthermore, suppose the region is divided into two distinct layers, the lower having a strong superadiabatic gradient and the upper being stably stratified. If the superadiabatic gradient in the lower layer is sufficiently large, the fluid will be convectively unstable (see Section 3.3.2), and motions will develop. When a steady convective state has been achieved, two significant features will be noticed. Firstly, the level at which the mean stratification changes from unstable to stable will have migrated upwards. Once convection arises the total flux in the lower layer will increase. Correspondingly, the flux in the upper layer must also increase, and this can be accomplished only by increasing the temperature gradient, i.e. by shrinking the physical extent of the upper layer (provided the boundary temperatures are held constant). This phenomenon is known as *penetration*. It has little relevance to constant-flux systems such as stars.

The second feature is far more important for our purposes. The convective motions do not cease at the level where the mean superadiabatic gradient vanishes: they extend up into the stably stratified region. Here there are no convective driving forces, so the motions must feed on any excess momentum and buoyancy present at the neutrally stratified level. This phenomenon is known as *overshoot*.

The physical processes occurring in the neighbourhood of the transition from a convectively unstable to a stable mean stratification, the so-called interfacial layer, are more varied than those in systems which are convective throughout. Because of the relevance of this physical situation to the interpretation of the granulation, we are fortunate that a simple laboratory experiment with an almost Boussinesq fluid can provide useful insight.

We are indebted to water for having the exceptional property of reaching maximum density not at its freezing point, but somewhat above it, at 4 °C. Thus if we maintain the upper surface of a water layer at room temperature and progressively cool the underside, a density inversion will be produced as the lower boundary passes through 4 °C. The resulting motion has been described by Myrup, Gross, Hoo & Goddard (1970) as 'upside-down convection'. In the lower part heat is *convected* downwards and in the upper stably stratified part heat is *conducted* downwards.

The onset of instability in such a system with a lower boundary at 0 °C was analysed by Veronis (1963), who defined an effective Rayleigh number $\mathcal{R}a_e$

$$\mathcal{R}a_e = g\Delta\rho d_3^3/(\rho_m \chi \nu), \tag{3.78}$$

where ρ_m is the maximum density of water (i.e. at 4 °C), $\Delta\rho$ is the density difference between 0 and 4 °C, and d_3 is the thickness of the unstable layer between 0 and 4 °C. Only d_3 can be varied in an experiment! Veronis found that the critical Rayleigh number for marginal stability decreases and the corresponding horizontal scale of the instability increases as the upper boundary temperature is increased, i.e. as the thickness of the stable regime is increased and d_3 correspondingly decreased. He also found that motion develops before the critical Rayleigh number is reached through the intervention of a finite-amplitude instability.

The motion of the system at marginally supercritical Rayleigh numbers was studied by Musman (1968) using mean-field equations (see Section 4.2.3). A principal cell forms which occupies the whole depth of the unstable region and overshoots into the stable layer. The degree of overshoot increases with the Rayleigh number and the cell becomes elongated vertically. The horizontal streaming within the interfacial layer at the top of the principal cell develops velocities comparable in magnitude to the largest vertical velocities in the unstable region. At low Rayleigh numbers, up to three countercells appear in the stable region above the principal cell, but they do not exert a significant effect on the dynamics of the main cell. As the Rayleigh number is increased, these countercells flatten, weaken, and disappear.

Musman considered only steady flows; at still higher Rayleigh numbers, one would expect unsteady motions to appear. The development of the motion to this stage has been investigated numerically by Moore & Weiss (1973b). They solved the full non-linear equations in a two-dimensional model. At low Rayleigh number ($\mathcal{R}a_e < 5\,\mathcal{R}a_{ec}$) the solutions were similar to those of Musman but predicted less penetration than the mean-field solutions. At $\mathcal{R}a_e \sim 28\,\mathcal{R}a_{ec}$, steady solutions could no longer be found. Instead, there was a persistent finite-amplitude oscillation, most marked in the interfacial layer.

To understand the origin of these oscillations we must recall the discussion of buoyancy forces in Section 3.3.1, where it was noted that in a stably stratified medium a displaced element will be subject to a restoring force, i.e. a negative buoyancy force. We can thus expect a wave-like motion to develop, as can be demonstrated explicitly with the (incompressible) Boussinesq equations (3.38–40). Consider small-amplitude adiabatic motion in an isothermal medium with a constant density scale height H_ρ. Then

$$\beta = (T_0/\rho_0)\,\mathrm{d}\rho_0/\mathrm{d}x_3 = -\,T_0/H_\rho < 0 \tag{3.79}$$

and, taking $\chi = 0$, the linearized equations yield an equation of motion

$$\partial^2 \nabla^2 v_3' / \partial t^2 = -(g/H_\rho)(\partial^2/\partial x_1^2 + \partial^2/\partial x_2^2) v_3'$$
$$= -N^2(\partial^2/\partial x_1^2 + \partial^2/\partial x_2^2) v_3', \qquad (3.80)$$

where N is the Brunt-Väisälä frequency.

This equation admits progressive wave solutions of the form

$$v_3' = v_{03}' \exp(\mathrm{i}\omega t) \exp(\mathrm{i}\mathbf{k} \cdot \mathbf{x}), \qquad (3.81)$$

where ω and $\mathbf{k}$ obey the dispersion relation

$$\omega^2 k^2 - N^2 k_\perp^2 = 0. \qquad (3.82)$$

Note that if $k_3 = 0$ all elements in a fluid column oscillate in phase at the Brunt-Väisälä frequency, as shown in Section 3.3.1.

The waves described by Eqns. 3.81 and 3.82 are known as internal gravity waves and are discussed by, for example, Townsend (1966). Bray & Loughhead (1974: Section 6.3) describe the internal wave modes of an isothermal compressible atmosphere, and Mihalas & Toomre (1981) those of a model of the solar atmosphere.

Moore & Weiss explain their results in terms of a non-linear interaction between convective oscillations and gravity waves. Circulating blobs of abnormally cold and hot material alternately raise and lower the extent of the overshoot, thus causing an undulation of the boundary between the convective cell and the stably stratified fluid. If the period of this oscillation matches an internal gravity mode in the stable region a resonance will occur. However, this phenomenon is likely to be an artifact created by the choice of convenient boundary conditions that imply an endless repetition of the cellular structure in the horizontal plane. In unbounded systems such resonances would not develop and certainly could not persist if the flow were made stochastic by further increase of the Rayleigh number.

We are thus led to the topic of turbulent overshoot. This phenomenon is extremely complicated, but we are again fortunate that many laboratory experiments have been conducted in this regime (Townsend, 1964; Myrup *et al.*, 1970; Adrian, 1975). The results deserve close attention and, in particular, comparison with the polytropic solutions for fixed boundaries described in Sections 4.3.2 and 4.3.3.

Let us start with the mean stratification. Fig. 3.6(*a*) shows the mean temperature and density profiles measured by Adrian (1975), whose experiments were the most carefully controlled and monitored. Throughout most of the convective region the stratification is almost neutral. In fact, the region is slightly stable, having a temperature just below 4 °C and a negative density gradient.

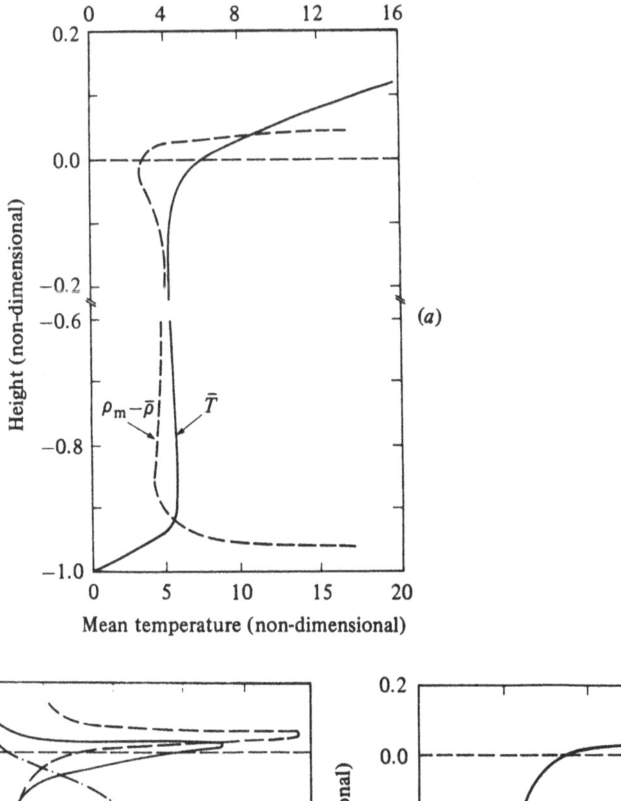

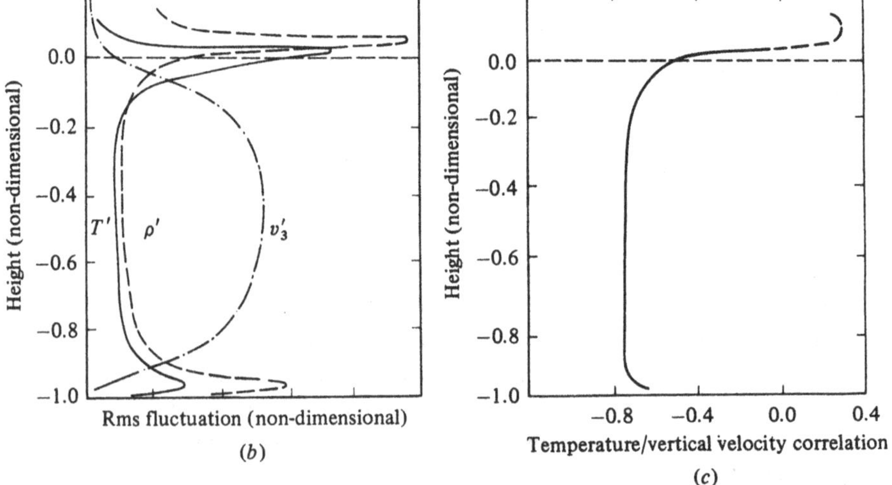

Fig. 3.6. The ice–water convective system (Adrian, 1975).
(*a*) Mean temperature and density as a function of depth. The zero line indicates the level at which water reaches its highest density ($\bar{T} = 3.98\,°\mathrm{C}$) and marks the limit of the zone of convective instability. Plotted is the difference between the maximum density and the mean density.

Only in the conduction-dominated layer at the bottom of the system and in the interfacial layer is there strongly unstable stratification. This is very similar to the behaviour of an isolated polytrope.

The scaled rms fluctuations of temperature, density, and vertical velocity are shown in Fig. 3.6(*b*). The temperature and density (buoyancy) variations peak very strongly in the conduction layer and in the interfacial layer *just above the level of neutral stratification*, the maximum density variations occurring somewhat further into the stable region than the maximum temperature variations. The vertical velocities are very noticeably damped in the stable region.

The convective heat flux is shown by the temperature/vertical velocity correlation in Fig. 3.6(*c*). This is generally negative, lying between −0.7 and −0.8, but rises rapidly throughout the interfacial layer. The positive values in the stable region imply a slight reverse convective flux (i.e. in the opposite direction to the conductive flux), but are of doubtful validity.

Adrian did not measure the horizontal component of the velocity, but Townsend (1964) noted that horizontal flow with considerable shear dominated the motion of tracers near the top of the constant temperature region and just above it (the interfacial region). Townsend further suggested that the large temperature and density fluctuations there were due to internal gravity waves excited by the impact of the overshooting turbulent convective elements. This suggestion finds some confirmation in Myrup *et al.* (1970), who observe that the fluctuations in the stable layer are occasionally seen to be 'directly correlated' with upward-moving convective elements below. Townsend (1964, 1966) analysed theoretically the excitation of gravity waves and demonstrated that they would be contained within a narrow layer above the overshoot region if the density stratification were smooth and the exciting (convective) impulse were of long duration. The wave packet then tends to spread horizontally rather than vertically.

These experimental results fall quite naturally into the picture of developed convection drawn in Sections 3.4.3 and 3.4.4. Moreover, they clearly demonstrate the occurrence of coherent large-scale structures in turbulent convection. These first become apparent in the third-order moments, the skewness of the temperature and vertical velocity fluctuations (see Section 2.3.4). In a normal distribution the skewness is zero, but here we encounter distributions with, on

Caption for Fig. 3.6 *contd.*

(*b*) Rms fluctuations in temperature, density and vertical velocity as functions of depth.

(*c*) Temperature/vertical velocity correlation coefficient as a function of depth.

one side, a widely extended tail. These large-amplitude excursions are uncommon but are nevertheless highly significant. They account for 75–80% of the total temperature variance and, being strongly correlated with velocity, they account for 95% of the heat flux in the convective region.

When the flow pattern is made visible, by dye-streaks for example, these correlated flows appear as long-lived, relatively narrow plumes or columns extending from top to bottom of the convective layer. The lifetime implies that these are not just the result of a random superposition of different scales but are definite entities. The coherent structures persist however turbulent the system is made. The incoherent fluid, whilst forming the bulk of the convective layer, is essentially passive: both the heat transport and the structural character of the convection can be quite accurately described by the coherent structures alone.

Close to the bottom of the convective layer, cold rising columns are predominant. These are the result of instabilities in the narrow conduction zone at the very bottom of the layer and drain away the local build-up of buoyant fluid.

In the upper half of the convective zone, warm downward columns are more common (see Fig. 3.7(*d*)). This has no natural explanation in terms of a boundary layer. Instead, the formation of downdrafts seems linked with the

Fig. 3.7. Measured frequency distribution of the temperature excursions at four scaled depths $\tilde{x}_3$ in the interfacial region of the ice–water experiment. The distributions change from being highly skewed, to bimodal, and then skewed in the opposite sense (Adrian, 1975).

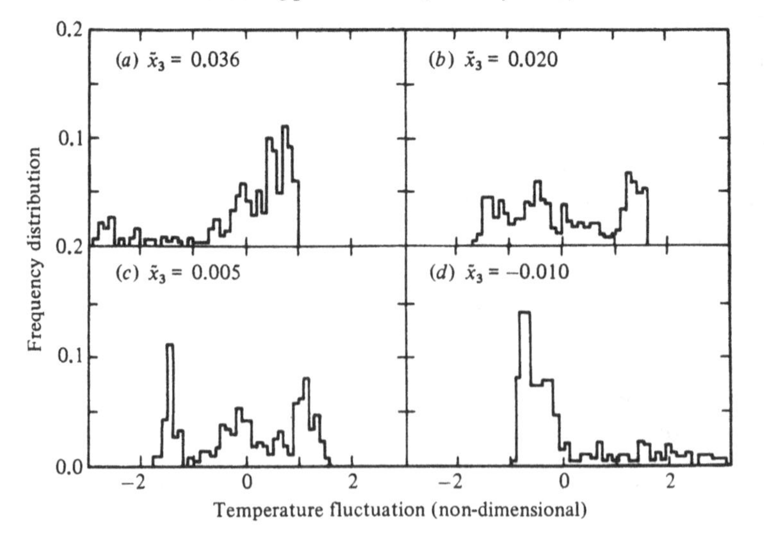

difficult problem of entrainment, by which weakly turbulent fluid from the stably stratified regions is mixed with highly rotational fluid in the convective layer (see Section 3.4.4).

The temperature fluctuation histograms for the overshooting region are shown in Figs. 3.7(*a*), (*b*), (*c*). Above the region of maximum temperature fluctuation (Fig. 3.7(*a*)), small positive temperature excursions due to waves are most frequent; the rarer large negative contributions are generally associated with rising material and presumably represent convective overshoot. The generally low correlation between temperature and velocity ensures that the convective heat flux is negligible. In the region of maximum fluctuation (Figs. 3.7(*b*), (*c*)), a bimodal temperature distribution is found, indicating the presence of two distinct streams, a cool upward stream from the convective layer and a warm downward stream from the stable layer. The former is again overshoot; the latter is surmised to be the 'breaking' of gravity waves.

Large-amplitude gravity waves can propagate with little damping in water but will suffer non-linear distortion. This causes the crest to arch over so that high density fluid finds itself above low density fluid. This configuration is unstable and the wave 'breaks', creating a patch of turbulence. Such limited patches of turbulence are well known in atmospheric physics. The downward breaking of the wave could trigger the downward convective plume referred to above. Or it may be that simple subsidence of the overshooting material is sufficient. In either case, the overturning of gravity waves appears to be a reality and is, in fact, revealed in the flow visualizations of Townsend (1964).

The implications of the 'ice-water' experiment are far-reaching. The presence of the coherent structures described above justifies the use of truncated modal expansions or numerical modelling with a coarse spatial grid. Non-linear effects must be incorporated, but the passive role of the small scales suggests that the particular model used to characterize their influence on the larger scales is unlikely to be important. More critical in convection with free surfaces is the nature of the interactions at the upper and lower boundaries. If the coherent structures are simply plumes travelling up, impinging on the upper boundary, subsiding and generating a downward plume, a coarse-grid model will provide a consistent representation of the whole process. If, on the other hand, the overshooting material diffuses into the stable layer, thereby generating gravity waves and dissipating its turbulence, the downward plumes require breaking gravity waves to entrain the irrotational fluid into the convective layer; an essential part of the physics will then be missing from a coarse-grid representation.

It must be stressed that the conditions in the solar atmosphere differ from those applying in the ice-water experiment. In particular, the efficient radiative energy transfer in the optically thin regions cannot be ignored. It is easy to retain a conductive term in Eqn. 3.80 and thus derive the modified dispersion relation

$$\omega^2 k^2 - N^2 k_\perp^2 = i\omega k^4 \chi. \tag{3.83}$$

This does not admit solutions with real ω and $\mathbf{k}$. Instead it gives radiatively damped oscillations: in response to a given impulsive disturbance, the waves generated will disperse and decay. The velocity damping is given by the imaginary part of ω

$$\mathscr{I}m\ \omega = k^2\chi/2, \tag{3.84}$$

for small damping ($4k_\perp^2 N^2 > k^6\chi^2$). The larger the radiative conductivity, the more rapid the damping.

In the low photosphere, the damping is so large (see Section 4.4.4) that the waves can hardly propagate. Furthermore, the amplitude of stochastically excited waves cannot be expected to build up by superposition, so large-amplitude internal waves and their associated non-linear effects are unlikely to occur. The processes by which the turbulence in the overshooting motions decays and the irrotational fluid is again incorporated into the convective layer of the Sun remain to be clarified.

3.5 Concluding remarks

We have introduced the general theory of convection in this chapter by successively isolating various terms in the equations and reducing them to their simplest forms. We have then tried to understand their physical content and to interpret them as competing components of the more complicated systems that occur in Nature.

We saw how an imposed temperature gradient leads to convection when the buoyancy force generated by a slight disturbance overcomes the drag due to atomic viscosity. The rotational character of the ensuing motion is described by the vorticity. Thereafter, the pattern of the convective motions is controlled by essentially two parameters which measure the magnitude of the imposed temperature gradient ($\mathscr{R}a$) - or, alternatively, the heat flux that the system is required to carry ($\mathscr{N}u$) - and the relative rates at which the convective motions lose kinetic energy through viscous and inertial effects ($\mathscr{P}r$). In low Prandtl number fluids the energy is lost mainly through non-linear inertial effects which produce an energy cascade to smaller scales and lead to turbulent motion. At the boundaries of a convectively unstable layer these motions overshoot, generating waves in the surrounding material before they are ultimately damped.

These processes have been illustrated by describing laboratory experiments, which are limited almost exclusively to non-stratified Boussinesq fluids with a particularly simple form of heat conduction. Nevertheless, all the effects described are present in stratified fluids, as we have demonstrated in some detail in our discussion of the onset of convection. In the next chapter we shall examine the theory of fully developed convection in stratified fluids - astrophysical convection.

Additional note

The conclusion that convection cells of supergranule size are not influenced by solar rotation has been challenged by D. H. Hathaway (1982, *Solar Physics*, **77**, 341-56). P. A. Gilman & G. A. Glatzmaier (1981, *Astrophysical Journal, Supplement Series*, **45**, 335-49) have formulated the anelastic equations for convection in a compressible rotating spherical shell required for a rigorous investigation of this problem (Section 3.2.3).

4

THE THEORY OF ASTROPHYSICAL CONVECTION

4.1 Introduction

In Chapter 3 we examined in some detail the physical processes occurring in convection at various stages of development and their interplay in Boussinesq laboratory fluids. We did not, however, attempt a theoretical description of developed convection in stratified systems - our ultimate concern.

We begin this task in Section 4.2, where we describe an approximate model that is basically Boussinesq and bypasses the formal solution of the hydrodynamic equations (3.18–27). In the upper part of the solar convection zone $\mathscr{R}a$ is very large and $\mathscr{P}r$ very small (Section 4.2.2). On the basis of laboratory experience one would expect the solar plasma to be highly turbulent, a topic to be discussed further in Section 5.3. Historically, the recognition of the role of stochastic processes in convection led to the analogy with the kinetic theory of gases which underlies the 'mixing-length' theory of convection introduced by the German aerodynamicist L. Prandtl in the early 1930s. Although this theory ignores virtually all details of the hydrodynamic processes, it has found wide application in the study of solar and stellar convection zones simply because it is hard to improve upon. Its ubiquitous acceptance should not blind us to the inconsistency inherent in the use of a Boussinesq theory to describe non-Boussinesq systems. It does, however, permit a zeroth-order description of the solar convection upon which more refined treatments can be based.

Attempts to extend the mixing-length model to include eddy dynamics and non-local effects are described briefly in Sections 4.2.3 and 4.2.4, but the *ad hoc* manner in which they are incorporated cannot be properly justified. Stationary, finite-amplitude convection, as observed on the Sun, requires us to seek a correct balance between non-linear and non-local effects.

Before turning to more rigorous formulations, we examine (Section 4.3.1) a more plausible application of the mixing-length model, in the description of the small-scale turbulent component of the velocity spectrum. This serves to introduce the concepts of turbulent viscosity and diffusivity, which represent the effects of small-scale motions on the large-scale flow.

We then abandon the Boussinesq approximation as a basis for a description of astrophysical convection and turn in Section 4.3.2 to the next level of approximation, the anelastic. This allows both density stratification and non-linear effects to be incorporated.

However, owing to mathematical difficulties, the development of this theory is still in its infancy, and only very limited use has so far been made of it. We describe two applications of the single-mode anelastic approximation: the first is to a polytropic atmosphere - the archetypal stratified system - and the second to a more realistic model of a stellar atmosphere, that of an A-type star. Such calculations reveal many details of the convective pattern, e.g. cell shape, stream-lines and isotherms. But the cell size is not determined. It must be chosen either arbitrarily or by imposing some condition, for instance, that the heat flux be maximized.

Such constraints are unnecessary when the full equations are solved, as some numerical simulations attempt to do. The techniques necessary for this are still under development. Of the computations reported in Section 4.3.3 only the two-dimensional ones can be relied on, and then only up to moderate Rayleigh numbers (100 $\mathscr{R}a_c$) and down to moderately small Prandtl numbers (0.1).

Finally, in Section 4.4, we move to a discussion of radiative energy exchange in the uppermost layers of the convection zone. An adequate treatment of this problem is essential to an understanding of the granulation. At present, however, there is a regrettable gap between the worlds of hydrodynamics and radiative transfer. In an effort to bridge this gap, we include in Sections 4.4.1 and 4.4.2 some fundamentals of radiative transfer theory of particular relevance to the problem of interpreting the solar granulation. The derivation and validity of the commonly employed approximations are discussed in Section 4.4.3. With this insight we then examine the strongly non-local smoothing effects caused by horizontal radiative transfer in the granulation layers (Section 4.4.4).

The material in this chapter has been presented with the aim of confronting dynamical models of the granulation with our observational knowledge wherever possible. Reviews of wider scope have been published by Spiegel (1971, 1972), with emphasis on physical principles and widely drawn illustrative examples. The subject of stellar convection has also been well reviewed by Gough (1977b).

A list of the principal symbols employed in this chapter will be found in Chapter 3, Table 3.1.

4.2 Mixing-length convective theories

4.2.1 Introduction

The mixing-length theory provides a link between laboratory experience and astrophysical models. It is based on a conceptual model suggested by

Boussinesq fluid experiments but does not pretend to be a self-consistent formulation for convection in stratified fluids. Nevertheless, when properly calibrated, the simple theory outlined in Section 4.2.2 yields a plausible model of the mean structure – the run of mean temperature and pressure – of the solar convection zone. It does so by attempting to evaluate adequately only the *mean* convective heat flux, and it can thus be regarded as providing a zeroth-order model. The simplicity of the formulation has also led to its almost exclusive use in calculations of stellar models (Cox & Giuli, 1968).

Its apparent success and popularity has encouraged several authors to take the model more seriously and to investigate the dynamics of the convective eddies in the spirit of the mixing-length concept. Various models have been produced based on different physical pictures. Although their descriptions of the eddy dynamics also vary (Section 4.2.3), they retain one common feature, viz. free parameters whose values are often adopted from laboratory experiment. They are hence firmly rooted in Boussinesq experience.

Early experiments demonstrated that turbulent motions transmit stresses to the mean flow parallel to a wall far beyond the regions in which viscous forces are effective. An empirical law for the gradient of the mean velocity perpendicular to the wall is

$$\partial \bar{v}_1/\partial x_3 = (\sigma_R/\rho)^{1/2}/(0.4x_3) = (-\overline{v_1'v_3'})^{1/2}/(0.4x_3), \tag{4.1}$$

where x_3 is the distance from the wall and σ_R is the Reynolds stress tensor (Section 4.3.1). Noteworthy in this formula is the combination of local and non-local quantities, a characteristic feature of mixing-length theories. The shear is determined partly by the local kinetic energy of the turbulence and partly by the distance of the point from the boundary, clearly a non-local quantity.

The mixing-length model takes its name from the hypothesis, due to Prandtl (see Section 1.4), that turbulent eddies can be treated like molecules interacting through intermittent collisions. Kinetic theory yields an equation of the same form as Eqn. 4.1, but with the factor $0.4x_3$ replaced by the mean free path between collisions l.

Such a model has certain deficiencies: in reality, the eddies interact continuously and there is no clear separation between the scale of the turbulence and that of the mean flow. Eqn. 4.1 with $l = 0.4x_3$ is better regarded as *defining* a characteristic length that provides a consistent scaling when the length scales of the mean flow and turbulence are the same.

Turbulent *convection* reveals no empirical scaling law, such as Eqn. 4.1, on which to base a model. In mixing-length theory, instead of solving the equations for the small-scale turbulent motions, we simply impose an *ad hoc* model on them. The motion is imagined to be broken up into characteristic eddies, generated by large-scale convective instabilities, each moving independently. The exact dynamical model is of less importance than the notion that the eddies rise or fall a characteristic distance between the 'collisions' in which they lose their identity. For example, it is immaterial whether the eddy is assumed to lose its momentum instantaneously after travelling a distance l or whether it loses its momentum continuously, since an object of size l subject to turbulent drag is brought to rest in a distance of $\sim l$ (Gough, 1977a). Likewise, it is of little consequence whether heat is conserved during the motion, as envisaged by Prandtl, or whether it is transferred continuously, as assumed by Öpik (1950). The eddies are regarded as the only effective means of transporting heat. Smaller scales resulting from turbulent cascade are taken to be uncorrelated with thermodynamic parameters, and thus to make no net contribution to the heat flux.

These suppositions form the basis of mixing-length models.

4.2.2 *Simple mixing-length theories*

The turbulent eddy picture suggests a different decomposition of the physical variables from that introduced in Section 3.2.2. The magnitudes of the various parameters in highly turbulent systems may deviate markedly from those in any initially static configuration, but observations suggest that the fluctuations about the average values are relatively small. Under these circumstances, it is convenient to express all physical quantities in the form

$$X = \bar{X} + X', \tag{4.2}$$

where $\bar{X}$ is some mean value, usually the average across a horizontal plane. The assumptions underlying the mixing-length model imply that the scale of the motion – the mixing length l – is smaller than the scale of the total convecting system. Moreover, most mixing-length theories implicitly assume that the Boussinesq equations adequately describe the fluctuations.

The governing equations for the mean quantities are then

$$\mathrm{d}(\bar{p} + \bar{p}_{\mathrm{c}})/\mathrm{d}x_3 = \mathrm{d}(\bar{p} + \bar{\rho}\, \overline{v_3'^2})/\mathrm{d}x_3 = -g\bar{\rho} \tag{4.3}$$

$$\mathrm{d}(\bar{F}_3 + \bar{F}_{\mathrm{c}})/\mathrm{d}x_3 = \mathrm{d}(\bar{F}_3 + \bar{\rho}\, C_p \overline{v_3' T'})/\mathrm{d}x_3 = 0, \tag{4.4}$$

and for the fluctuations

$$\nabla \cdot \mathbf{v}' = 0 \tag{4.5}$$

$$\partial \mathbf{v}'/\partial t + (\mathbf{v}' \cdot \nabla)\mathbf{v}' - \overline{(\mathbf{v}' \cdot \nabla)\mathbf{v}'} = -\nabla p'/\bar{\rho} - \delta T' \mathbf{g}/\bar{T} + \nu \nabla^2 \mathbf{v}' \tag{4.6}$$

$$\partial T'/\partial t - \beta v_3' + (\mathbf{v}' \cdot \nabla) T' - \overline{(\mathbf{v}' \cdot \nabla) T'} = -(1/C_p \bar{\rho}) \nabla \cdot \mathbf{F}'. \tag{4.7}$$

The convective motions give rise to an effective convective pressure $\bar{p}_{\mathrm{c}}$ in the hydrostatic support equation (4.3) and to a convective heat flux $\bar{\mathbf{F}}_{\mathrm{c}}$ in the heat transport equation (4.4).

If we consider only the larger (convectively driven) scales we may ignore the viscous terms (see Section 3.4). Furthermore, the Boussinesq assumptions allow us to eliminate the pressure fluctuations (see Section 3.2.3). The equation for the vertical motion then becomes

$$\partial\nabla^2 v_3'/\partial t + \nabla^2[(\mathbf{v}'\cdot\nabla)v_3' - \overline{(\mathbf{v}'\cdot\nabla)v_3'}] - \partial\nabla\cdot[(\mathbf{v}'\cdot\nabla)\mathbf{v}' - \overline{(\mathbf{v}'\cdot\nabla)\mathbf{v}'}]/\partial x_3$$
$$= (\delta g/\bar{T})(\partial^2/\partial x_1^2 + \partial^2/\partial x_2^2)T', \quad (4.8)$$

while the energy equation, including conductive heat transport, is

$$\partial T'/\partial t - \beta v_3' + (\mathbf{v}'\cdot\nabla)T' - \overline{(\mathbf{v}'\cdot\nabla)T'} = \chi\nabla^2 T'. \quad (4.9)$$

We are interested only in the effects of the convection on the mean equations, represented by the dynamic (convective) pressure and convective heat flux; in the simplest form of the theory we do not need to follow the detailed motion of the individual eddies.

We shall follow essentially the treatment given in the review article by Gough (1977a). Formulae with just one free parameter, the mixing length l, can be obtained from Eqns. 4.8-9 by simple dimensional analysis. If an eddy starts from rest under the influence of buoyancy, the non-linear and pressure forces can initially be neglected and, taking $\partial/\partial t \sim v_3'\partial/\partial x_3$ and integrating along the path, Eqn. 4.8 becomes

$$v_3'^2 \simeq g\delta T' x_3/\bar{T}. \quad (4.10)$$

As a typical velocity we may take the value at the midpoint of the eddy's free path ($l/2$), i.e.

$$\hat{v}_3'^2 \simeq g\delta \hat{T}' l/(2\bar{T}). \quad (4.11)$$

Likewise the energy equation (4.9) can be integrated approximately to give

$$T' - \beta x_3 \simeq -\chi T'/(2v_3' x_3), \quad (4.12)$$

and a typical temperature fluctuation $\hat{T}'$ is then

$$\hat{T}' - \beta l/2 \simeq -\chi\hat{T}'/(l\hat{v}_3'). \quad (4.13)$$

These equations can be solved for the typical velocity and temperature fluctuations, and the convective heat flux can be evaluated in the same spirit, giving

$$\bar{F}_c = \bar{\rho} C_p \overline{v_3' T'} \simeq \bar{\rho} C_p \hat{v}_3' \hat{T}'$$
$$= \mathscr{S}^{-1}[(1+\mathscr{S})^{1/2} - 1]^3 K\beta/4, \quad (4.14)$$

where $\mathscr{S} = \delta\beta g l^4/(\chi^2\bar{T})$.

The dynamic pressure can be estimated similarly, but is usually negligible in stellar convection.

Since we have neglected the viscosity, $\mathscr{R}a$ and $\mathscr{P}r$ cannot appear explicitly in the expression for the heat flux. However, the quantity $\mathscr{S}$ will be recognized as the product of the Prandtl number and a Rayleigh number based not on the total dimension of the system but on the mixing length l:

$$\mathcal{S} = \mathcal{R}a \cdot \mathcal{P}r, \tag{4.15}$$

where

$$\begin{aligned}\mathcal{R}a &= (\delta/\bar{T})(\beta g l^4/\chi\nu) \\ &= (\delta/\bar{T})\,\beta g l^4 (\bar{\rho} C_p/K)(\bar{\rho}/\eta).\end{aligned} \tag{4.16}$$

$\mathcal{S}$ alone determines the nature of the turbulent convection in the sense that it measures its efficiency.

If $\mathcal{S} \gg 1$, the heat flux is

$$\bar{F}_c \sim K\beta\,\mathcal{S}^{1/2}/4. \tag{4.17}$$

In this case the superadiabatic gradient required to carry a given heat flux is small and decreases slowly with increasing convective efficiency. The convective velocities likewise decrease.

On the other hand, if $\mathcal{S} \ll 1$, the heat flux becomes

$$\bar{F}_c \sim K\beta\,\mathcal{S}^2/32. \tag{4.18}$$

The superadiabatic gradient then required to provide a given heat flux is large and increases rapidly with decreasing efficiency. The convective velocities increase correspondingly.

Other formulations - Öpik (1950), Böhm-Vitense (1958), Cox & Giuli (1968) - yield basically similar expressions, differing only in their numerical coefficients. Gough & Weiss (1976) have calibrated the Böhm-Vitense and Öpik formulae by evolving a solar model to the Sun's present age and adjusting the mixing length until the model luminosity and radius matched the observed solar values. This procedure allows the initial helium composition of the Sun to be determined, but the initial proportion of heavy metals (Z) remains a free parameter. For a range of metal abundances the two mixing-length prescriptions yield almost identical convection zone structures; those for $Z = 0.02$ are shown in Fig. 4.1.

Throughout most of the convection zone the various ionization processes (H I, He I, He II) reduce the adiabatic gradient to the value at which the stratification becomes superadiabatic. However, the radiative conductivity is very low so that even small superadiabatic gradients generate very large values of $\mathcal{R}a$: the convection is efficient.

Fig. 4.1 also shows that the convection zone extends down far below the level of the ionization zones. This has a simple explanation. The need for heat to go into ionization energy holds down the mean temperature rise but the pressure increases with depth due to the rapidly accumulating weight of overlying material. Once the ionization is complete, the temperature can rise more rapidly but cannot immediately achieve the *T-p* conditions for convective stability. Accordingly, the temperature gradient in the transition zone remains superadiabatic and convection continues.

We can now examine whether the inaccessible layers fulfil the conditions for turbulent convection.

If we were to follow Gough, Moore, Spiegel & Weiss (1976) we would evaluate $\mathscr{R}a$ and $\mathscr{P}r$ at the midpoint of the convection zone, $x_3 \sim -100\,000$ km. However, the deepest point for which the relevant model parameters are available is 15 000 km (Spruit, 1974). Here we find

$$\beta \simeq 10^{-6}\ \mathrm{K\,m^{-1}}$$

$$\bar{\rho} \simeq 2\ \mathrm{kg\,m^{-3}}$$

$$\bar{T} \simeq 10^{5}\ \mathrm{K}$$

$$\bar{p} \simeq 2.5 \times 10^{9}\ \mathrm{N\,m^{-2}}$$

Fig. 4.1. Structure of the solar convection zone predicted by two mixing-length models, showing temperature T, logarithmic superadiabatic gradient $\nabla - \nabla_{ad}$, and the fraction of heat flux carried by convection F_c/F. The solid curves and the lower of the two pressure scales (above) were computed according to Böhm-Vitense, the dashed curves and upper pressure scale according to Opik (see text). The zones of partial hydrogen and helium ionization (10 to 90%) are indicated by arrows (Gough & Weiss, 1976).

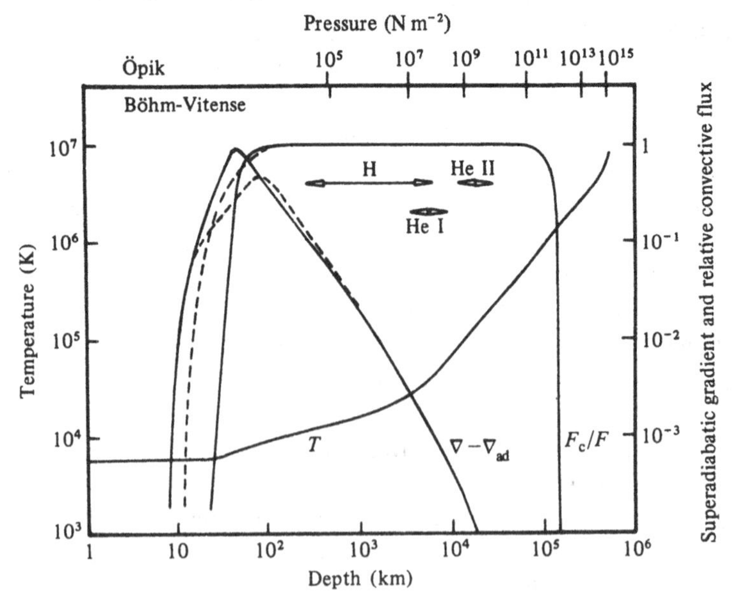

$$C_p \simeq 4 \times 10^4 \text{ m}^2 \text{ s}^{-2} \text{ K}^{-1}$$

$$\chi \simeq 2.6 \text{ m}^2 \text{ s}^{-1}$$

$$\nu \simeq 1 \times 10^{-4} \text{ m}^2 \text{ s}^{-1}$$

$$d \simeq 2 \times 10^8 \text{ m},$$

where we have taken the viscosity from Edmonds (1957). Substitution yields $\mathscr{P}_{\!r} \sim 4 \times 10^{-5}$ and a Rayleigh number based on the total depth of the convection zone of $\sim 10^{28}$. The last quantity is so enormous that uncertainties in the choice of parameters and the depth at which they are evaluated are of no importance. The conditions for the onset of turbulent convection are thus overwhelmingly satisfied (see Section 3.4.4).

To determine the strength of the turbulent convection, we must evaluate the local Rayleigh number. Taking a scale of $\sim$10 000 km in the interior, we obtain

$$\mathscr{R}a \sim 10^{23}, \quad \mathscr{S} \sim 10^{18},$$

implying very efficient convection, as noted above.

At the upper boundary of the convection zone, near the visible surface of the Sun, the situation is more complicated. As one proceeds downwards the sharp rise in opacity before the onset of hydrogen ionization reduces the effectiveness of radiative conduction and causes the stratification to become superadiabatic. On the other hand, the low densities result in a lower $\mathscr{R}a$ and less efficient convection. This is borne out by the values of the parameters already given (Section 3.4.4) for the base of the visible layers, calculated using a local scale height of 250 km:

$$\mathscr{P}_{\!r} \sim 10^{-9}, \quad \mathscr{R}a \sim 2 \times 10^{11}, \quad \mathscr{S} \sim 2 \times 10^2.$$

This (relatively) less efficient convection requires a larger superadiabatic gradient and larger convective velocities; moreover, the structure of the layer, some 1000 km thick, is sensitive to the details of the convective model employed (Fig. 4.1).

Below this layer most models converge to the same adiabat. Thus a *single* consistent model determines the interior structure, the depth of the convection zone, and its properties throughout the adiabatic layers. These values are essentially model-independent.

Mullan (1971) has claimed that the solar convection zone is only 10 000 km thick rather than some 150 000-200 000 km as required by Gough & Weiss, but the discrepancy can be explained by Mullan's failure to fit his convection zone model to the correct interior model.

Spruit (1974), who employed the standard interior model, also predicted a deep convection zone. He introduced no less than three further free parameters multiplying the right-hand sides of the expressions 4.11, 4.13 and 4.14. He determined these 'efficiency' parameters by comparing model predictions with the centre-to-limb brightness curves calculated using the semi-empirical Harvard–Smithsonian Reference Atmosphere (see Section 5.2.2). Since these adjustable parameters are given no physical basis, no further light is thrown on the convective processes; the results cannot be generalized to produce stellar models. The insight required to do this must be based on a better understanding of the convective dynamics.

4.2.3 Eddy dynamics

The simple mixing-length theories described in the previous section have very limited aims. They set out to provide a consistent prescription for calculating the mean structure of a stratified convective system based on Boussinesq behaviour. No attempt is made to produce realistic dynamical predictions that can be compared with measurements of the fluctuations of temperature and velocity.

It is possible to do this, to some extent, by modelling the competing processes of buoyancy, advection, and mass loss and gain by the *eddy*. The manner in which these processes are incorporated tends to depend more on some preconceived physical analogy than on mathematical analysis. A brief account of these developments is nonetheless of interest, not only for seeing how far the mixing-length concept may be pursued, but also as providing an introduction to the apparently differing visualizations of the dynamics in terms of eddies and thermals. We shall meet both eddy and thermal prototypes in this chapter, and we shall attempt a resolution in Section 5.3.

As an example, we follow Gough (1977a), who describes a model based on an earlier one due to Speigel (1963). They start from the premise that eddies form at random throughout the fluid but each can be regarded as a cell fixed in location throughout its growth. They further assume that the cell size remains smaller than the scale of the mean atmospheric stratification and that non-linear effects are not important for most of the lifetime of the eddy. Then the governing perturbation equations, 4.8 and 4.9, become

$$\partial \nabla^2 v_3'/\partial t = (g\delta/\bar{T})(\partial^2/\partial x_1^2 + \partial^2/\partial x_2^2)T' \tag{4.19}$$

$$\partial T'/\partial t - \beta v_3' = \chi \nabla^2 T'. \tag{4.20}$$

If small-scale motions are included by means of an eddy 'viscosity' and 'diffusivity' (cf. Section 4.3.1), the additional terms are also linear ($+\nu_t \nabla^4 v_3'$

and $+\chi_t \nabla^2 T'$ are added to the right-hand sides of Eqns. 4.19 and 4.20 respectively).

The set of linearized perturbation equations plus the equations for the mean quantities are known as the mean-field equations. The mean-field equations are non-linear since they include a quadratic term involving the product of the fluctuating temperature and vertical velocity and, when solved self-consistently, lead to true finite-amplitude solutions. However, if we attempt to solve the perturbation equations independently of those for the mean flow, the non-linearities are lost and the problem is not completely specified. But this is precisely what mixing-length models set out to do.

Let us suppose that the infant cell (eddy) is locally described in terms of normal modes of the form

$$X' = X'_0 \exp(nt) \exp(\mathrm{i}\mathbf{k} \cdot \mathbf{x}), \tag{4.21}$$

where $l = \pi/k_3$ is taken to be the vertical scale. We do not require that the normal modes represent the fluctuations *throughout* the convection zone at any one time, which Hart (1973) has demonstrated to be untenable.

Substituting Eqn. 4.21 into 4.19 and 4.20 we obtain growing and damped modes whose growth rates n are

$$n_\pm = \Psi^{-1} \mathscr{S}^{-1/2} [\pm(1 + \Psi^2 \mathscr{S})^{1/2} - 1](g\delta\beta/\bar{T}\Phi)^{1/2}, \tag{4.22}$$

where

$$\Psi = 2\pi^{-2} \Phi^{-3/2} (\Phi - 1) \tag{4.23}$$

$$\Phi = 1 + k_3^2/k_\perp^2. \tag{4.24}$$

Moreover, the ratio of temperature to velocity amplitude is

$$T'_0/v'_{03} = n_\pm \Phi \bar{T}/g\delta. \tag{4.25}$$

Φ and Ψ are geometric factors dependent on the ratio of the vertical to the horizontal size of the cell. As in Section 4.2.2, $\mathscr{S}$ can be interpreted as the product of a Prandtl number and a Rayleigh number based on the local mixing length l. Estimates of neither Φ nor l are given by this analysis and they remain free parameters throughout.

Only the eddies which grow exponentially with time interest us. After a time t, the vertical velocity will be

$$v'_3(t) = v'_{03} \exp(n_+ t), \tag{4.26}$$

and the fluid will have moved a vertical distance

$$\Delta x'_3(t) = v'_{03} [\exp(n_+ t) - 1]/n_+. \tag{4.27}$$

Gough now makes the *ad hoc* assumption that the eddies are disrupted by shear forces when the velocities have grown sufficiently. He assumes that the probability of dissolution is proportional to the instantaneous shear; the survival probability, P_s, then follows a Poisson distribution

$$P_s \sim \exp(-\Delta x_3'(t)/l). \tag{4.28}$$

In order to apply linear dynamics we have assumed that the eddy growth Θ during the time $\hat{t}$ taken to travel the mean free path l is large, i.e.

$$\Theta = \exp(n_+\hat{t}) \gg 1, \tag{4.29}$$

where

$$l = \Delta x_3'(\hat{t}) \sim v_{03}' \exp(n_+\hat{t})/n_+. \tag{4.30}$$

Making the assumption that, at any time t, the convective layer is completely filled by eddies, we then know how many eddies were born at a previous time t_0 and have since acquired a temperature excess $T'(t-t_0)$ and a vertical velocity $v_3'(t-t_0)$. Averaging over the eddy distribution yields the mean convective flux

$$\bar{F}_c = \tfrac{1}{2}\Phi^{-1/2}\Psi^{-3}\mathscr{S}^{-1}[(1+\Psi^2\mathscr{S})^{1/2}-1]^3 K\beta/(\ln\Theta - 0.577). \tag{4.31}$$

The dynamic pressure can be evaluated similarly:

$$\bar{p}_c = \tfrac{1}{2}\bar{\rho} l^2 n_+^2/(\ln\Theta - 0.577). \tag{4.32}$$

These expressions contain the geometric factors Φ, Ψ and the eddy growth Θ as well as the mixing length l; $\bar{F}_c$ and $\bar{p}_c$ are only weakly dependent on the eddy growth. More important are the choices of the free parameters Φ and l. Similar free parameters arise in other models of the eddy dynamics.

Ulrich (1970a) postulates a different set of equations: following Moore (1967) he visualizes convection as an ensemble of atmospheric thermals based on Turner's (1964) model. The weakness of this description is again the treatment of the non-linear processes of dissolution and coalescence to form new thermals. Turner's model, which rests on the concept of a spherical 'Hill vortex', is empirical in nature, the parameters being derived from experiment. If these are incorporated, the modifications due to Ulrich introduce only one major unknown, the initial mean size of the thermal. This parameter corresponds to the mixing length of other formulations.

Although Ulrich's model is basically Boussinesq, the size of the thermal is allowed to vary as the density of the medium changes and can be large compared with the scale height, unlike the normal mixing length in the eddy approach.

In general, there is no satisfactory way of determining *a priori* the mixing length in the various formulations. Arguments have been put forward, based on linear theory, for $1 \leqslant \Phi \leqslant 3$. Choosing the marginally stable mode, for instance, requires $\Phi = 3$. Böhm-Vitense's (1958) formulae imply $\Phi = 2$.

Attempts over the years to arrive at a satisfactory physical interpretation of l have made little progress. It has been suggested that l describes the largest possible convective scale, but it is not clear what this is in free convection. Some early authors argued that, in the absence of containing surfaces, the stratification provided the cutoff; the expansion or contraction experienced by an eddy moving through a scale height would be sufficient to destroy its identity. Thus the mixing length should be of the order of the density scale height. In order to avoid numerical difficulties associated with density inversions in stellar atmospheres (see Section 5.2.4), Böhm-Vitense assumed the mixing length to be related to the *pressure* scale height, which is always positive. If the mixing length is then written

$$l = \alpha H_p \quad \text{or} \quad l = \min(|x_3 + \Delta x_3|, H_p), \tag{4.33}$$

where $x_3 + \Delta x_3$ measures the extent of the travel close to the surface, the unknown parameter α (assumed constant), or Δx_3, can be determined by comparison with solar models (Spruit, 1974; Gough & Weiss, 1976).

The mean stratification predicted by various models of the convection zone is always very similar (see Section 4.2.2). The basic assumptions and hence functional form of the heat flux are common to all mixing-length models. The derived values of l vary since different authors describe the small-scale eddy dynamics differently. For example, Gough & Weiss (1976) have compared the models of Öpik (1950) and Böhm-Vitense (1958) and find that the values of α, the ratio of mixing length to scale height, are respectively 2.4 and 1.1.

The basic physical inconsistency of the mixing-length theory is now obvious. The only way of defining a scale in an unbounded fluid is to identify it with the distance over which the fluid properties change. But the calibrations require the mixing length to be at least comparable with this scale, contradicting the premise that the motions are 'small-scale'. A more detailed treatment within the Boussinesq approximation does not lead to a satisfactory description of the eddy dynamics. There is little point in further refining such *ad hoc* models by incorporating a rough treatment of Reynolds stresses (Gough, 1977a), or by trying to take account of the small effects arising from the coupling between rotation and convection (Durney & Spruit, 1979).

4.2.4 *Non-local mixing-length theories*

The mixing-length theories described so far are local theories in the sense that the convective flux at any given point is proportional to the mean superadiabatic gradient β at that point. It is assumed that β does not vary significantly over one scale height in the atmosphere, and that β steadily approaches zero at the boundaries and the convective velocities and flux follow suit. However, in the Sun the convective velocities are observed to extend beyond the point where β vanishes. This behaviour is intimately related to the presence of large variations of β over distances of a scale height in the outermost layers of the solar convection zone (see Fig. 4.1). To accommodate these effects, the assumption of constant β must be relaxed to produce a non-local theory.

The simplest non-local theory is that proposed by Spiegel (1963). He noted that, for variable β, the solution of the linearized equations for the eddy growth (Eqns. 4.19–20) is still given by Eqn. 4.22, but with the local value of β replaced by the averaged value

$$\bar{\beta}(x_3) = \frac{\int \beta v_3'^2 \, dx_3'}{\int v_3'^2 \, dx_3'} = \frac{2}{l} \int_{x_3 - l/2}^{x_3 + l/2} \beta(x_3') \cos^2[\pi(x_3' - x_3)/l] \, dx_3'. \tag{4.34}$$

In this case, the contributions to the heat flux and turbulent pressure from eddies centred at different heights must be taken into account. If the local flux (Eqn. 4.31) is denoted by $\bar{F}_c^*$, the non-local value is given by the integral

$$\bar{F}_c = \frac{2}{l} \int_{x_3 - l/2}^{x_3 + l/2} \bar{F}_c^*(x_3') K(x_3, x_3') \, dx_3', \tag{4.35}$$

where the kernel is

$$K(x_3, x_3') = \cos^2[\pi(x_3' - x_3)/l]. \tag{4.36}$$

The local flux $\bar{F}_c^*$ must be evaluated with the non-locally averaged superadiabatic gradient $\bar{\beta}$.

Other models of the turbulent eddy motion due to Spiegel (1963) and Ulrich (1970b) lead to different expressions for $\bar{F}_c^*$ and to different kernels, since the averaging procedure depends on whether the elements are envisaged to be cells, rising and falling bubbles, or thermals.

The mathematical difficulties of solving a combination of integral and differential equations are considerable and have prompted the choice of other kernels that would allow the integral equations to be reformulated as differential equations. But this is to tailor the physics to mathematical convenience (Travis & Matsushima, 1973; Ulrich, 1976; Nordlund, 1976).

Non-local theories may be calibrated in exactly the same manner as local theories (see Section 4.2.2). Models incorporating eddy dynamics, such as the Spiegel eddy model or the Ulrich thermal model, have more free parameters than local theories. In principle, the choice of parameters can be checked, e.g. by comparison with the centre-to-limb variation of the photospheric brightness fluctuation. This would allow a detailed comparison of the *dynamical* predictions with observations. Such a procedure has yet to be carried through, although Travis & Matsushima (1973) and Ulrich (1970b) have reported comparisons restricted to simplified versions of these models having but one free parameter. These latter two formulations yield almost identical results (Nordlund, 1974).

A second group of non-local theories, comprising the models of Parsons (1969) and variants of Ulrich's model due to Nordlund (1974), is based on a different premise. The previous theories, whilst admitting a variable super-adiabatic gradient, maintain the other fluid variables constant over the eddy scale. Nordlund points out that this assumption fails most markedly for the radiative diffusivity at the upper boundary of the convection zone. There the opacity scale height drops to 60 km, and the layers become optically thin extremely rapidly. Any temperature excess of the convective element is then quickly radiated away, and the temperature fluctuations are effectively destroyed.

Nordlund argues that on the Sun the motions should overshoot the limit of convective instability but carry no excess heat. To reproduce this, in the second group of models the velocity is derived from non-local considerations but the convective flux is estimated locally. Then the decay of the convective flux at the top of the convection zone is as rapid as in local mixing-length theories.

Nordlund's argument, however, exaggerates the effects of radiative cooling at the surface. If the excess heat supply extends down into the optically thick regions, the cooling rate is not a local quantity but some average over the whole region (see Section 4.4.4). Since the local cooling rate decreases sharply in the optically thick layers, the actual mean rate in the atmosphere is lower than local estimates would indicate, and the temperature perturbations are not destroyed so effectively. Evidently the true convective flux must lie between the values given by the local and non-local values (see Section 5.3).

Keeping within the framework of mixing-length theories the treatment of the non-local thermal effects can thus be improved. What cannot be improved, however, is the *ad hoc* treatment of non-linear processes. Yet it is just the balance between non-linear and non-local effects which determines the scale of the motion in turbulent convection (Section 4.3.2). By their very nature Boussinesq theories fail in this respect.

It is worth pausing here to repeat the reasons why the theory, despite its internal inconsistency, is apparently successful in describing the *mean* structure of the solar convection zone.

Firstly, the aims of the theory are limited. It is designed, and is generally used, to provide only an estimate of the local convective heat flux. Moreover, this estimate contains a free parameter that is calibrated by requiring the global parameters of the Sun to be correctly matched. This ensures that, on average, the heat flux is correctly given by the prescription despite the fact that the functional form of the heat flux (measured by the Nusselt number) predicted by Eqn. 4.17,

$$\mathcal{N}_u - 1 \propto (\mathcal{R}_a \cdot \mathcal{P}_r)^{1/2}, \qquad (4.37)$$

is supported by neither laboratory nor numerical experiment. Whether, under such circumstances, the calibration can be validly assumed to hold for other stars is unknown.

Secondly, and perhaps more surprisingly, there is a general agreement between the models of the solar convection zone predicted by the various formulations of the mixing-length theory. This arises simply because the convection is efficient throughout most of the zone and the true temperature gradient is very close to the adiabatic gradient. Differences arise only in the outermost 1000 km where the convection is inefficient. The influence of this layer is not sufficient to affect the overall structure, and the layer simply adjusts, without visible effect, to the need to transport the heat flux incident from below. The situation is quite different for stars with extensive zones of inefficient convection, such as red giants, and the mean structure derived on the basis of mixing-length theories must then be regarded with suspicion.

To return to the Sun, the mixing-length model of the solar convection zone provides, by virtue of its unique calibration, a perfectly adequate zeroth-order approximation to the run of mean temperature and pressure. The importance of this fact should not be underestimated. It demonstrates that under solar circumstances we must concentrate on turbulent convection. Furthermore, it tells us that over most of the solar convection zone the superadiabatic gradient and the convective velocities are very small. The dynamical effects of these layers on the surface are likely to be negligible. We should instead focus on those layers where the gas experiences significant acceleration, beginning at the depth of 5000 km at which He II and H II start to recombine. Here we require a self-consistent theory of turbulent convection in a stratified medium.

One procedure for improving the description is to treat the mixing-length solution as a starting point for an iterative solution. It can be used to solve the equations for the fluctuating quantities, whose values can then be used to solve the equations for an improved estimate of the mean structure, and so on. As a procedure, this cannot be faulted, but as yet the iteration has been carried out only as far as the linearized first-order equations.

Böhm (1963a,b) analysed the growth rates and velocity eigenfunctions of both fundamental and harmonic solutions to the fluctuation equations, neglecting turbulent transport processes (an inconsistency). Since the background atmosphere is assumed to be convectively unstable, it is no surprise that exponentially growing modes are found. However, all the usual objections to linearized treatments apply - they do not indicate the relative importance of the modes, nor do they yield the streamlines, cell evolution and lifetime. Just as in the modelling of eddy dynamics, no observational quantities are predicted; instead the observations must be used to make a selection from the limitless number of possible solutions.

Much the same criticism can be levelled at the work of Roxburgh & Tavakol (1979). These authors performed a stability analysis using the mean-field equations (Section 4.2.3). The analysis is formally analogous to that of the onset of convection in a Boussinesq fluid but with a turbulent viscosity replacing the atomic viscosity and conductivity (see Section 4.3.1). Instability was found for supercritical turbulent Rayleigh numbers ($\mathscr{R}a_t = \delta\beta g d^4/(T\nu_t^2)$). Again this is little more than a consistency test since the equations were generated on the assumption of convective instability. We reiterate that one cannot simply identify the critical modes with the modes that finally dominate in steady finite-amplitude convection.

Higher-order non-linear theories require a model of the small-scale motions, and here the mixing-length description makes a more respectable appearance. We shall consider this aspect in Section 4.3.1 before dealing with anelastic convective theories in detail in Section 4.3.2.

4.3 Generalized theories of astrophysical convection

4.3.1 Turbulent viscosity and diffusivity

We now proceed to the next level of approximation, namely the anelastic and other models more rigorous than the mixing-length model. Again, however, we are faced with a basic, although different, inconsistency which deserves attention at the outset.

In the following we shall seek 'exact' solutions to the hydrodynamic equations that describe the flow pattern in detail. At first sight this may appear impossible as we have already concluded in Section 4.2.2 that the flow in the solar convection zone is turbulent. However, although this dilemma has no mathematically satisfactory resolution, a physically plausible solution can be found based on our earlier description of turbulent convection (Section 3.4.4). The small-scale motion, many orders of magnitude smaller than the convective scales, is much more random and can be realistically described by a fully stochastic model. It is thus convenient to treat the turbulence separately, ignoring the influence of convection, and to then incorporate the turbulent terms in the equations for the large-scale ordered convective flow.

This is not strictly legitimate; it supposes a separation of turbulent and convective components that does not exist in reality since the convective and turbulent scales merge. Nevertheless, the dichotomy is attractive in practice as it replaces one extremely complicated process, turbulent convection, by two weakly coupled simpler processes, convection and turbulence.

A further advantage is the more legitimate use of mixing-length estimates for the parameters of the turbulent component since the scales are, by this stratagem, much smaller than those of the mean stratification. So let us evaluate the effects of these scales before turning (in Section 4.3.2) to the large-scale motion.

We follow Wasiutynski (1946: Chapter 1, Section 2) and Unno (1969) and average all quantities over the smallest convective scale, so that the turbulence appears as a fluctuation about a local mean

$$X' \to X' + X''. \tag{4.38}$$

The convective quantities X' are now in fact mean quantities but we shall not overload the notation in order to distinguish them as such. Furthermore, we shall assume that the stochastic nature of the turbulent component X'' allows us to replace the spatial means by ensemble averages denoted by $\bar{X}''$. With the decomposition (Eqn. 4.38), the convective momentum equation becomes in the anelastic approximation (cf. Eqn. 3.24)

$$\begin{aligned} &\partial(\rho_0 v_i')/\partial t + \rho_0 v_k' \, \partial v_i'/\partial x_k \\ &\quad = -\,\partial p'/\partial x_i - g\rho' \delta_{i3} - \partial(\rho_0 \overline{v_k'' v_i''})/\partial x_k. \end{aligned} \tag{4.39}$$

On the scale of the convective flow the viscous terms are negligible and do not appear in Eqn. 4.39. With the aid of the relation (cf. Eqn. 3.23)

$$\partial(\rho_0 v_k'')/\partial x_k = 0, \tag{4.40}$$

the mean fluctuating inertial term has been rewritten in a form which emphasizes its role in replacing the viscous stress. In fact, the turbulent momentum flux,

$$(\sigma_{\mathrm{R}})_{ki} = -\rho_0 \overline{v_k'' v_i''}, \tag{4.41}$$

is known as the Reynolds stress tensor (see Section 4.2.1). This term is a turbulent drag force; it removes kinetic energy from the mean convective flow and feeds it into the turbulent motion.

For a simple gas ($\delta = 1$) the convective energy equation is obtained from Eqn. 3.25 in the form

$$\begin{aligned} &\rho_0 C_p \, \partial T'/\partial t - \partial p'/\partial t - C_p \beta \rho_0 v_3' - C_p \beta \rho_0 T' v_3'/T_0 \\ &\quad = -\,\partial F_i'/\partial x_i - T_0 \, \partial(\rho_0 \overline{v_k'' s''})/\partial x_k + C_p \beta \rho_0 \overline{T'' v_3''}/T_0. \end{aligned} \tag{4.42}$$

If we assume that buoyancy plays no significant role in the turbulence dynamics (Section 3.4.4) the final (anisotropic) term may be discarded. The turbulence is then seen to affect the mean heat transport solely through its entropy flux.

We thus need estimates of the turbulent momentum and entropy fluxes in terms of the mean quantities, which the mixing-length approach allows us to do. We have already seen how closely the mixing-length theories are based on analogy with molecular processes (Section 4.2.1), so this treatment of turbulent motion leads naturally to a turbulent viscosity and diffusivity that closely resemble, but are additional to, the molecular transport coefficients.

To see this, let us follow the motion of a turbulent element from the point $\mathbf{x}_0$ at time t_0. Bearing in mind that the turbulence dynamics are decoupled from the larger scales, the velocity of the element relative to the mean flow after time t can be written to first order

$$v_i''(\mathbf{x}, t_0+t) = v_i''(\mathbf{x}_0, t_0) + tX_i'' - tv_k''\partial v_i'/\partial x_k, \tag{4.43}$$

where X_i'' is the turbulent force component (Unno, 1969). The entropy of the element is then

$$s''(\mathbf{x}, t_0+t) = s''(\mathbf{x}_0, t_0) + tY'' - tv_k''\partial s'/\partial x_k, \tag{4.44}$$

where

$$Y'' = (-\nabla\cdot\mathbf{F}/\rho T)''. \tag{4.45}$$

The last terms on the right-hand sides of Eqns. 4.43 and 4.44 make allowance for the separation of the turbulent element from its notional starting point in the mean flow after a time interval t.

Let us suppose that the turbulent velocity components are initially uncorrelated with one another and with s'', $\mathbf{X}''$, Y'' and that after travelling a certain distance (not necessarily the same for momentum and entropy) the element merges again with its surroundings and loses its correlations. Multiplying both Eqns. 4.43 and 4.44 by $\rho_0 v_k''$ and taking the ensemble average at the travel times t_0+t_m, t_0+t_s for the momentum and entropy respectively, we obtain

$$\rho_0\overline{v_i''v_k''} = \rho_0\overline{v_{i0}''^2}\delta_{ik} - \rho_0 t_m\overline{v_{i0}''^2}\,\partial v_k'/\partial x_i - \rho_0 t_m\overline{v_{k0}''^2}\,\partial v_i'/\partial x_k \tag{4.46}$$

$$\rho_0\overline{v_k''s''} = -\rho_0 t_s\overline{v_{k0}''^2}\,\partial s'/\partial x_k, \tag{4.47}$$

where the summation convention no longer applies. The subscript zero on certain averages indicates that these terms should be evaluated at the beginning of the motion; they will be replaced by typical values.

In keeping with our assumption of isotropic turbulence, Eqns. 4.46 and 4.47 can be written

$$\rho_0\overline{v_i''v_k''} = p_t\delta_{ik} - \rho_0\nu_t(\partial v_k'/\partial x_i + \partial v_i'/\partial x_k) \tag{4.48}$$

$$\rho_0\overline{v_k''s''} = -\rho_0\chi_t\,\partial s'/\partial x_k. \tag{4.49}$$

If $t_m = t_s$, the turbulent diffusivity χ_t is equal to the turbulent or 'eddy' viscosity ν_t.

The turbulent effects are thus reduced to a turbulent pressure p_t and modified viscous and diffusive terms as a result of our mixing-length approach. Similar expressions have been derived rigorously from the hydrodynamic equations for the case of homogeneous isotropic turbulence by Krause & Rüdiger (1974) and Eschrich (1978). The more general conditions under which vector (momentum) and scalar (entropy) quantities can be taken to diffuse in a similar manner have been defined by Knobloch (1977).

The above treatment yields the general form of the turbulent terms but does not allow us to evaluate them since t_m, t_s and $\overline{v''^2}$ remain undetermined. Estimates are generally obtained partly by experiment and partly by sleight of hand. Laboratory experiments measure the effective Reynolds and Péclet numbers defined as

$$\mathcal{R}e_t = \langle v'' \rangle l / \nu_t \simeq 1/(\hat{k} \langle v'' \rangle t_m) \tag{4.50}$$

$$\mathcal{P}e_t = \langle v'' \rangle l / \chi_t \simeq 1/(\hat{k} \langle v'' \rangle t_s), \tag{4.51}$$

so that

$$\nu_t \simeq \langle v'' \rangle / (\mathcal{R}e_t \hat{k}) \tag{4.52}$$

$$\chi_t \simeq \langle v'' \rangle / (\mathcal{P}e_t \hat{k}) \tag{4.53}$$

where $\langle v'' \rangle$ is the ensemble rms velocity and $l \sim 1/\hat{k}$ is the characteristic size of the turbulent eddy.

According to Unno (1969), empirical values of $\mathcal{R}e_t$ are soundly based, but those of $\mathcal{P}e_t$ are not. He takes provisionally

$$\mathcal{R}e_t \sim 10, \qquad \mathcal{P}e_t \sim 5.$$

These values yield a turbulent Prandtl number

$$\mathcal{P}r_t = \mathcal{P}e_t / \mathcal{R}e_t \sim 0.5.$$

It is commonly assumed that this ratio is unity in turbulent fluids, implying that the diffusion of entropy and momentum by turbulence can be treated alike; but this is not supported by experiment.

The empirical data still leave the mean scale and rms velocity of the turbulence undetermined. This is where inconsistency arises: since the larger turbulent scales are energetically dominant most authors have been tempted to substitute the convective scale and velocity for the turbulent quantities. However, this makes nonsense of the procedure of separating the turbulent and convective scales described above.

Nevertheless, we have reduced the effect of the turbulent scales on the flow equations to an enhanced viscosity and diffusivity, and we can now proceed to examine the solution of the resulting non-linear equations.

4.3.2 Anelastic convective theories

A glance at the anelastic equations derived in Section 3.2.3 indicates that there is little hope of deriving analytic formulae for the convective heat flux. Indeed, all anelastic results reported so far have been obtained numerically. They require extremely sophisticated numerical techniques, including careful attention to the propagation of numerical errors. Since we are interested only in the hydrodynamic results, these computational difficulties will be glossed over. Full accounts are to be found in the literature cited.

The anelastic equations presented in Section 3.2.3 employ a decomposition about a *static* state. A slightly different form arises when the quantities are expressed as fluctuations about a mean (Gough, 1969; Latour, Spiegel, Toomre & Zahn, 1976). The general approach to the solution of these equations has already been touched upon in our discussion of the linearized equations (Section 3.3). The cellular character of the granulation pattern suggests that the horizontal configuration be specified in some suitable harmonic manner. If complete orthonormal solutions of the harmonic equation

$$\partial^2 f/\partial x_1^2 + \partial^2 f/\partial x_2^2 = -k_\perp^2 f \tag{4.54}$$

are taken as a basic set, corresponding to all the independent linear modes, the cell planform of all quantities except the horizontal velocities can be expressed as

$$X = \iint X_{\mathbf{k}_\perp}(t, x_3) f_{\mathbf{k}_\perp}(x_1, x_2)\, \mathrm{d}\mathbf{k}_\perp. \tag{4.55a}$$

An integral rather than a sum is required as the spectrum of $\mathbf{k}_\perp$ is infinite and continuous for an unbounded medium. In order to satisfy the continuity equation (3.23), we must then allow the horizontal velocity components to have the form

$$v_j' = \iint v_{j\mathbf{k}_\perp}'(\partial f_{\mathbf{k}_\perp}/\partial x_j)\, \mathrm{d}\mathbf{k}_\perp \quad (j = 1, 2). \tag{4.55b}$$

The obvious next step is to discretize and truncate the expansion for X so that only a finite number of horizontal modes are included, i.e. a modal expansion. The subsequent steps depend on the method adopted for handling the stochastic variations in the large-scale flow.

Unno (1969), in the first anelastic model of convection, regarded the motion as basically stochastic but with coherent large-scale structuring. The instantaneous coherent structure, or 'eddy', he described in terms of a WKB approximation

$$X' \to X'(t)\langle X'(x_3)\rangle 2\cos[k_\perp x_1/2^{1/2} + \phi_{x_1}(t)]\cos[k_\perp x_2/2^{1/2} + \phi_{x_2}(t)]$$
$$\times \cos[k_3(x_3)x_3 + \phi_{x_3}(t) + \epsilon_X(x_3)], \tag{4.56}$$

where $X'(t)$, $\phi_{x_1}(t)$, $\phi_{x_2}(t)$, $\phi_{x_3}(t)$ are variables random in time.

By the assumption of stationarity, the ensemble-averaged equations are time-independent and provide a set of ordinary differential equations for the rms quantities $\langle X' \rangle$ and the relative phases ϵ_X.

Unno adopted the mixing-length estimates for the turbulent diffusivity (Section 4.3.1). He argued that the convective and turbulent spectra are continuous and cannot be distinguished; the proper estimates of the turbulent velocity and scale are just those of the representative convective mode. Thus only a single convective turbulence mode need be considered. Effective Reynolds and Péclet numbers were taken from experiment.

Since the equations are non-linear, the amplitudes and the scale of the convection are determined once a horizontal scale is chosen. Unfortunately, the equations are very complicated and seem to be unstable numerically (Travis, quoted in Travis & Matsushima, 1973) but limited numerical results have been obtained using a version of the equations due to Waters (van der Borght & Waters, 1971). However, these results have not been fully described.

An alternative approach (Waters, 1971; van der Borght, 1971, 1975a, b; Latour *et al.*, 1976), which has been extensively explored, treats the convective motion as a steady cellular flow with turbulent transport. The time derivatives in the equations for the fluctuations are then set equal to zero, and solutions are sought of the form

$$X' = \sum_{\mathbf{k}_\perp} X'_{\mathbf{k}_\perp}(x_3) f_{\mathbf{k}_\perp}(x_1, x_2). \tag{4.57}$$

The expansion is substituted into the anelastic equations. Separate, but coupled, equations for each mode can then be found by multiplying by each $f_{\mathbf{k}_\perp}$ and integrating over the horizontal plane, taking account of the orthogonality relation $\iint f_{\mathbf{k}_\perp} f_{\mathbf{k}'_\perp} \, dx_1 \, dx_2 = \delta(\mathbf{k}_\perp - \mathbf{k}'_\perp)$. This procedure is said to 'project out' the $\mathbf{k}_\perp$ mode.

Like Unno, van der Borght restricted consideration to one mode and approximated the effects of others by turbulent transport terms, taking a turbulent Prandtl number equal to unity (see Section 4.3.1). The resulting tenth-order system of ordinary differential equations for convection in a polytropic layer can be found in van der Borght (1975a). They were solved numerically with free boundaries over which the temperature fluctuations were assumed to vanish.

The horizontal structure is not uniquely determined by Eqn. 4.54. All solutions of the equation are equally appropriate but the choice of bimodal or rectangular cells

$$f = 2 \cos k_1 x_1 \cdot \cos k_2 x_2 \qquad (k_1^2 + k_2^2 = k_\perp^2) \tag{4.58}$$

leads to particularly simple equations. The cells are square for the case $k_1 = k_2 = k_\perp/2^{1/2}$. Van der Borght arbitrarily took the horizontal wavenumber to be $k_\perp = \pi/d_3$, where d_3 is the thickness of the convective layer.

The solutions demonstrate that, for low heat fluxes, the anelastic approximation is self-consistent; the size of the fluctuations increases towards the upper boundary but remains small everywhere. The flow takes the form of a single cell extending from the bottom to the top of the convecting polytrope; but no investigation of the influence of the choice of horizontal wavenumber on the organization of the flow has been carried out.

A more extensive investigation of the single-mode equations, which explored the effects of changing the horizontal planform and scale, was carried out by Toomre, Zahn, Latour & Spiegel (1976) for a model A-type stellar atmosphere. It was found that two-dimensional rolls produced large velocities and shears, and the authors argued that shear instabilities would then produce three-dimensional motion, a development excluded by the two-dimensional formulation. Thus roll solutions are unrealistic. On the other hand, hexagonal cells could transport the required heat flux without the 'strain' (large velocity gradients) apparently suffered by rolls.

The effect of the choice of planform is most marked in the advection terms in the equations for the fluctuating momentum and energy. When the modal equations are projected out these terms possess the coefficient

$$C = \tfrac{1}{2} \iint f_{\mathbf{k}_\perp}^3 \, \mathrm{d}x_1 \, \mathrm{d}x_2. \tag{4.59}$$

This quantity is known as the self-interaction coefficient and vanishes for two-dimensional rolls and the pseudo-two-dimensional flows considered by van der Borght. The elimination of the non-linear terms with coefficient C reduces the anelastic equations to a set of mean-field equations (see Section 4.2.3); only the temperature-velocity correlations remain. In the absence of advection, the buoyancy forces in a thin layer can be balanced only by viscous forces so, if the atomic viscosity is small, the velocities and shears are large. Van der Borght compensated for this effect by utilizing a Prandtl number of unity, i.e. a turbulent viscosity, thus reintroducing three-dimensional effects in a rough fashion.

Hexagonal planforms involve self-interaction; hence the buoyancy is balanced by the advective terms. This leads to smaller velocities without the need to introduce an artificially increased viscosity. The modelling in the two cases is thus essentially different, but it must be admitted that the results are very similar. Evidently, simply allowing for three-dimensional motion is of more importance than the particular planform chosen.

Toomre *et al.* (1976) also investigated the effects of changing the viscosity. They found it expedient to arbitrarily increase the viscosity above its atomic value in order to reduce transients and to prevent the formation of narrow boundary layers, but the three-dimensional solutions are very insensitive to the Prandtl number so long as it remains low. Provided the transfer of energy out of the convective scales is properly accounted for, it is unimportant how far the cascade must proceed before the viscous losses balance the vorticity production.

The calculations were performed with similar boundary conditions to those of van der Borght, but the procedure was significantly different. Instead of assuming a steady state, the time development of the system was followed explicitly until a stationary solution was achieved. The fundamental solutions were the same as those found by van der Borght and showed remarkable stability. Overtone solutions could be found but, when perturbed, always evolved into a fundamental in which the motion extended over the full depth of the unstable layer. No tendency was found for the motion to break up into cells stacked vertically.

The properties of hexagonal convection cells in extended polytropic atmospheres have been studied in some detail by Massaguer & Zahn (1980), who investigated both single- and two-horizontal-mode solutions. The most striking feature of their results is the significant role of the pressure fluctuations. In Boussinesq convection they play a very secondary role, being required only to drive the horizontal flow. In compressible convection this is not the case, as had been pointed out earlier by Hart (1973).

If we recall our estimates of the relative pressure fluctuations in Section 3.2.3,

$$\hat{p}'/p_0 \sim (d_3/H)(\hat{T}'/T_0) \sim (d_3/H)(\hat{\rho}'/\rho_0), \tag{4.60}$$

we see that if the motion extends over several scale heights the pressure fluctuations can no longer be neglected when considering the vertical motion. This is what occurs in fundamental cells. The pressure is required to increase at the top of the flow in order to spread the stream horizontally, and the size of the pressure fluctuation is large enough to dominate the temperature fluctuation and leads, through the equation of state, to a density *excess*, i.e. negative buoyancy. The vertical flow is thus braked by pressure forces and not by non-linear inertial forces. This effect will be compounded by efficient radiative cooling at the upper boundary, which tends to reduce the temperature fluctuations and their dynamical role still further.

The two-mode solution is particularly interesting since it allows the presence of cells with both upward and downward flow at the centre. The cell with upward motion at the centre was found to dominate the solution. Indeed, the total fluxes of enthalpy and mechanical energy were indistinguishable from those of the solution with but a single, upward mode.

However, the range of parameters covered by the calculations was very restricted, and an especially simple form of coupling between the modes was employed. Caution should be exercised in generalizing the results.

We shall conclude this section by comparing the single-mode results for polytropes obtained for square cells by van der Borght (1975a) with those for hexagonal cells by Massaguer & Zahn (1980). Fig. 4.2 shows the variation of the relative temperature and density fluctuations, vertical velocity and superadiabatic gradient for the two models. Unfortunately, the authors did not treat identical cases, so the model parameters are quite different. These parameters (see Section 3.3) are given in Table 4.1.

Van der Borght's model has a deeper, thicker layer (measured by $\tilde{x}_{03}$) with a much stronger stratification (measured by m). He also chose the horizontal wavenumber $\tilde{k}_{\perp}$ to be a compromise between the value of the critical wavenumber for the onset of convection and that which maximized the heat flux. Massaguer & Zahn opted for the critical value. The wavenumber that maximized the heat flux was 7.07 (recall from Section 3.3.3 that the critical wavenumber for the onset of convection is only 2.2!). The choice of $\mathscr{P}_{*}$ number has already been discussed in this section.

Since the variation of the mean quantities across the layer is less extreme in the model of Massaguer & Zahn, they could carry the computations to higher heat fluxes without numerical difficulties. In addition, they evaluated the corresponding $\mathscr{R}a$ at the midpoint of the layer; the model shown has $\mathscr{R}a = 10^4\,\mathscr{R}a_c$. Van der Borght provides no estimate of $\mathscr{R}a$, but it is clearly well in excess of the critical value.

Despite parameter differences and the very different treatment of the nonlinearities, the solutions are remarkably similar. In both cases, the mean stratification of the polytrope is adiabatic in the centre, with large superadiabatic gradients being confined to the boundary layers. The buoyancy is positive in the lower half of the convective layer, and accelerates the flow. The buoyancy is much greater in van der Borght's more highly stratified model, leading to larger vertical velocities, as measured by the vertical Mach number (the ratio of the vertical velocity to the local sound speed). The maximum velocities occur in the upper half of the layer where the densities are lower.

Despite the large temperature fluctuations and superadiabatic gradient near the surface, the pressure buildup causes a deceleration of the flow, most marked in the highly stratified model of van der Borght. Both examples clearly demonstrate the pressure excess by the presence of a strong negative buoyancy in the uppermost 25% of the layer.

Fig. 4.2. Two single-mode models of convection in polytropes (dashed line: van der Borght, 1975a; full line: Massaguer & Zahn, 1980). The various panels show the variation with height ($\tilde{x}_3 - \tilde{x}_{03}$) of the relative temperature fluctuation, relative density fluctuation, the vertical Mach number and mean superadiabatic gradient. For details of the models see text.

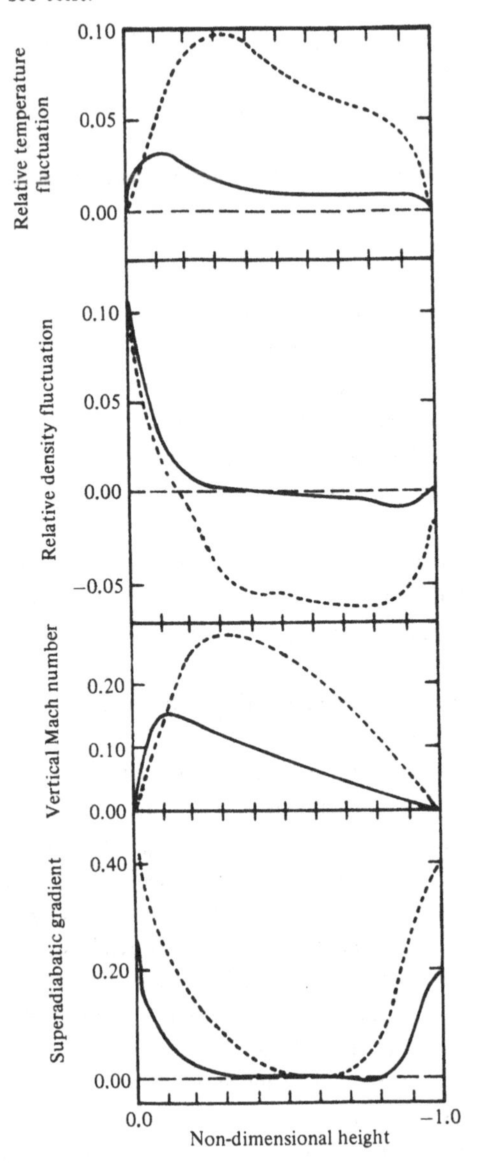

4.3.3 *Numerical modelling*

The simplified modal approximations to anelastic convection described above are attractive but the inevitable question arises: how well do they represent actual convection?

Experimental data is almost entirely lacking for stratified convecting systems, so the truncated models must be validated by comparing them with the results of accurate numerical modelling of the full hydrodynamic equations. However, the provision of a set of numerical solutions which could be used as a standard model would be a difficult and time-consuming task and has not yet been carried through systematically.

By far the most reliable and best documented work is that of E. Graham. He chose the most straightforward example of a stratified fluid, the polytrope.

In his earlier simulations, Graham (1975) considered two-dimensional compressible convection in a rectangular polytropic slab with free horizontal boundaries maintained at constant temperatures. The side walls were thermally insulating free boundaries. This problem has six free parameters, $\mathscr{R}a$, $\mathscr{P}r$, the polytropic index m, the aspect ratio of the slab (Γ = width/depth), the layer thickness $\tilde{x}_{03}$, and the ratio of specific heats γ. It should be noted that Graham measured the Rayleigh number at the top of the layer, whereas Gough *et al.* (1976) chose its midpoint. A grid of 21×21 points was employed and the time development was followed until a steady solution emerged. The same solution was reached regardless of the initial conditions.

The flow pattern for the solution $\mathscr{R}a = 10\,\mathscr{R}a_c$, $\mathscr{P}r = 1$, $m = 1.4$, $\Gamma = 1$, $\tilde{x}_{03} = 1$, $\gamma = \frac{5}{3}$, is shown in Fig. 4.3(*a*). In this case, the mean temperature and density drop by 50% across the layer and the pressure by 75%.

There is a marked asymmetry between the upward and downward flows. The downward flows have higher speeds and are hence more concentrated than the

Table 4.1. *Parameters of anelastic models of convection in polytropes*

	Model	
	van der Borght (1975a)	Massaguer & Zahn (1980)
$\mathscr{N}u$	~ 2	~ 10
$\mathscr{P}r$	1	0.01
γ	1.4	1.67
$k_\perp$	3.14	2.22
$\tilde{x}_{03}$	0.5	0.1
m	2	1

Note: for symbols, see Table 3.1.

upward. An explanation of this property in terms of the pressure effects has been given by Deupree (1976): the pressure rises where the upward flow impinges on the boundary and the pressure excess spreads across the cell, braking the upward flow and accelerating the downward flow. This behaviour has already been noted in the anelastic models (Section 4.3.2) and is an inevitable consequence of a flow extending across a layer in which stratification effects are significant.

Another noticeable feature is the approximate equality of the horizontal speeds near the upper and lower boundaries. At low $\mathscr{R}a$, this is due to a large kinematic viscosity near the surface, resulting from the assumption of constant dynamic viscosity. At higher $\mathscr{R}a$ ($\geqslant 10\,\mathscr{R}a_c$), it seems to be due to non-linear drag since even constant kinematic viscosity models show the same behaviour (Graham, 1977).

If the aspect ratio of the box is increased, rolls form side by side (Fig. 4.3(*b*)) each having a ratio of width to depth ($Q/2$) of about 2. There is some doubt as to whether a stationary solution was actually achieved in this case as a seemingly

Fig. 4.3. Numerical simulation of the two-dimensional velocity field in a polytrope confined within rectangular boxes of different aspect ratios. The stratification is not as strong, nor the Rayleigh number as high, as in the single-mode solutions of Fig. 4.2 (Graham, 1975).

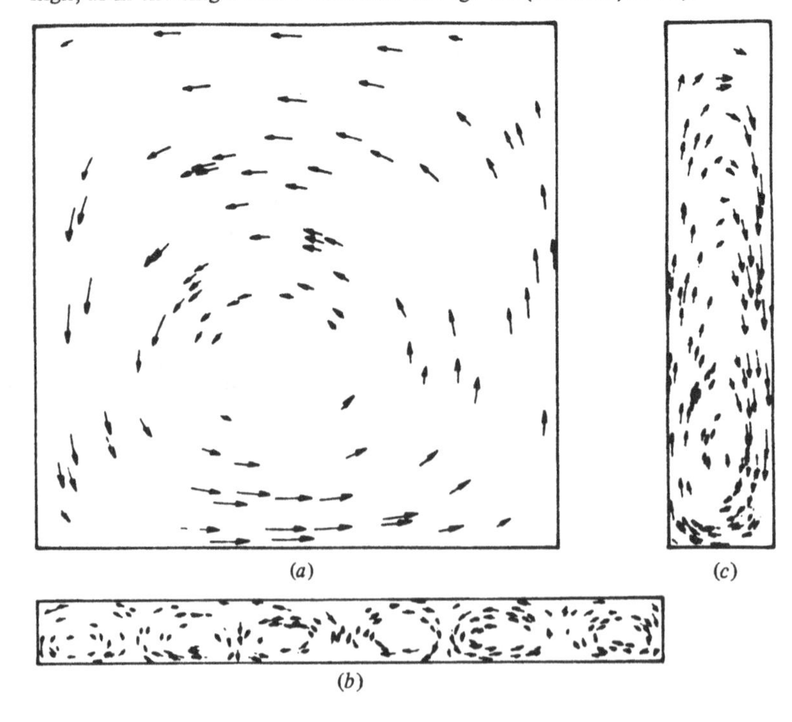

stable configuration of 7 rolls first developed and then suddenly switched to an equally persistent 6 roll configuration. The normalized horizontal wavenumbers were $\tilde{k}_\perp = 2.2, 1.9$ respectively; these values are equal to neither the critical wavenumber $\tilde{k}_{\perp c}$ nor the wavenumber that maximized the heat flux ($\tilde{k}_\perp \sim 3$). Decreasing the aspect ratio produced a single cell elongated vertically (Fig. 4.3(*c*)). Again, no tendency for the cells to break up into smaller vertical scales was found.

The computations were extended to a higher Rayleigh number ($100\,\mathcal{R}a_c$) and lower Prandtl number (0.1) without essential difference. As $\mathcal{P}r$ is reduced, the speeds increase sharply reaching a maximum Mach number of 0.5 at $\mathcal{P}r = 0.1$. But as noted above (Sections 3.4.3 and 4.3.2), this is a feature characteristic of two-dimensional flows and is unrealistic.

A more noteworthy result from the numerical simulations is the length of time required for the system to acquire a steady state. Increasing the aspect ratio increases the time. Decreasing the Prandtl number does likewise. The system 'winds up' slowly and we might suspect on the basis of Boussinesq results that, at low $\mathcal{P}r$, a three-dimensional instability would intervene before a steady flow is realized.

Graham (1977) has also reported briefly some three-dimensional simulations. Three-dimensional flows indeed fail to achieve a stationary state for modest $\mathcal{R}a$ if the horizontal extent of the container is made greater than the vertical dimension. Fig. 4.4 shows a realization of the velocity field for $\mathcal{R}a = 10\,\mathcal{R}a_c$ and $\tilde{x}_{03} = 1$. As in the Boussinesq case, vertical vorticity has been created, plainly visible in the horizontal swirling, and is responsible for the time-dependent character of the flow.

The models were not pursued into the fully turbulent regime. Decreasing the viscosity would decrease the scale at which viscous dissipation becomes important to the point at which the number of grid points required becomes prohibitively large. Exact three-dimensional low Prandtl number modelling seems out of the question with present-day computers.

Deupree (1975, 1976) chose to avoid this dilemma by modelling two-dimensional flows in a stratified fluid of vanishing viscosity ($\mathcal{R}a \to \infty$, $\mathcal{P}r \to 0$). His model has many attractive features. It is based on the envelope of an F-type main-sequence star and employs spherical geometry to describe a convectively unstable shell bounded by stably stratified layers. In the extreme case considered, the unstable layer extended over 20 000 km, i.e. three pressure scale heights. When the fluid was released from rest, cells first formed extending throughout the unstable layer and overshooting substantially into the stably stratified layer above. There seemed to be a preferred horizontal wavenumber, $\tilde{k}_\perp \sim 8$. Later the single cell broke up into two vertical cells each extending somewhat beyond one pressure scale height, but no stationary state was reached as no loss mechanisms were present. Deupree's simulations provide the *only* indication that convection at low $\mathscr{P}_*$ breaks up into eddies, as visualized by mixing-length theories. It thus behoves us to examine their basis more critically.

Two points then emerge. Firstly, it is not clear what happens to the horizontal vorticity produced in this system. In two-dimensional flow there can be no transfer to smaller scales and with vanishing viscosity there are no losses. The physics of a real fluid of low Prandtl number is thus not reproduced. Secondly, the grid was too small: in most of the calculations there were only 4×11 points within the convective layer. Weiss (1977) concluded that the 'resolution is too coarse for the results to be credible'.

4.3.4 Physical implications

Since the first edition of this book (1967) there has been an enormous growth in our understanding of convection. The problem has been attacked over a very wide front. The experiments have been refined, and almost every relevant mathematical tool has been mobilized. The heat flux has come under particularly intensive study, and there is general agreement between theory and laboratory experiment over a large range of Rayleigh and Prandtl numbers.

Fig. 4.4. Numerical simulation of the three-dimensional velocity field in a polytrope. The degree of stratification and Rayleigh number are the same as for Fig. 4.3. (The Prandtl number is not given in the original paper.) (Graham, 1977.)

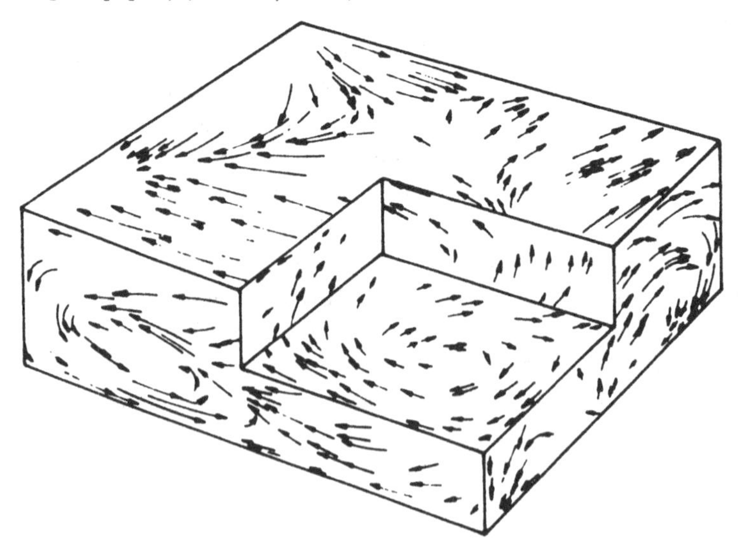

The results that concern us are almost unanimous in two findings: firstly, the mean stratification is almost everywhere neutral (adiabatic) and, secondly, the dominant scale of the motion extends from the top to the bottom of the unstable layer.

The first conclusion gives some support to proponents of mixing-length models, which also produce adiabatic stratification over most of the convection zone.

However, the second conclusion brings them little comfort. Horizontally extended layers break up into side-by-side cells, but an extended vertical layer does not form a hierarchy of stacked cells. The dominance of this pattern of motion means that the drastic simplification produced by restricting attention to the single global scale yields plausible results. When the motion extends over several scale heights, the pressure plays an important governing role in the dynamics – contrary to the case envisaged by the mixing-length model.

The processes which select the horizontal scale are not understood but, for moderate $\mathcal{P}_r$ fluids, experiment indicates that the flow configuration adjusts to maximize the heat flux and the horizontal scale adjusts itself to achieve this.

But how far is this true of astrophysical convection? Less progress has been made in answering this question. Non-mixing-length models of stellar convection zones are limited to those of Toomre *et al.* (1976), using the modal approximation, and of Deupree (1975, 1976), based on two-dimensional numerical simulations. There are no exact solutions with which these approximate solutions can be compared.

The more securely based findings of Toomre *et al.* are in accord with the general conclusions described above. If they can be carried over from an A-type star to a G-type star like the Sun, they imply that the dominant convective mode has a horizontal size comparable to the vertical extent of the whole convection zone. Yet, despite considerable efforts, giant convective cells with dimensions of the order of 200 000 km ($\frac{2}{7}R_\odot$) on the Sun have not so far been detected. If such shells do exist, their photospheric velocity amplitude must be $\lesssim 10\,\mathrm{m\,s^{-1}}$ (Howard & LaBonte, 1980).

Obviously, single-mode models are incapable of describing granulation, supergranulation, and the hypothetical giant cells. Such a treatment lies within the scope of multi-mode anelastic modelling, but the inclusion of several modes exacerbates the difficulty of an *a priori* choice of horizontal scales. Moreover, experience with low $\mathcal{P}_r$ fluids in the laboratory suggests that maximizing the heat flux may not be the appropriate criterion.

Multi-mode models lead directly to a further consideration. Toomre *et al.* found no difficulty in producing stationary solutions with just one mode, but it is by no means self-evident that this will continue to be so with two or more modes. The Lorenz equations (Lorenz, 1963) demonstrate the complexity inherent in just three non-linear equations - one velocity mode and two temperature modes. The equations have only two stationary solutions - one is trivial, the state of no motion, and the other corresponds to convective rolls. Above a certain critical parameter, analogous to the Rayleigh number, no stable stationary state appears to exist. Instead, the system changes continually in an erratic manner, lingering in the neighbourhood of a set of configurations known as 'strange attractors'. The stochastic nature of the motion has not been proven mathematically but is clearly indicated by numerical computations. A modern review of the mathematical and physical properties of non-linear dynamical systems is provided by Swinney & Gollub (1981).

These properties underlie the cell fragmentation process outlined by Moore (1979). He visualizes a stationary state developing in a low $\mathscr{P}_*$ fluid as a result of a balance between the competing processes of cell growth and fragmentation. A quasi-axisymmetric cell will slowly develop until the inertial terms grow large enough to dominate the viscous forces. An instability then results, triggered either by its own violation of axisymmetry or by a neighbouring disturbance, leading to fission. Each fragment is then a seed cell, so there is a marked continuity from generation to generation.

This is just another facet of the problem underlying the usage of the words 'cell' and 'thermal' by different authors. We are concerned with cells that do not develop a fully stationary flow, and it is a matter of semantics whether one terms the apparent structures cells or thermals. But if, on the basis of the terminology, the equations are tailored to describe different physical processes the distinction is not so innocuous. By the inclusion of multiple scales, anelastic modelling admits the possibility of the cells forming and dissolving and thus merges the two descriptions. This approach echoes the language of turbulence theory.

In this case, one works entirely in Fourier space and considers the wavenumber dependence of physical quantities, i.e. their spectra. Thus, for example, Ledoux, Schwarzschild & Spiegel (1961) and Eichler (1977) expand all quantities in terms of the three-dimensional wavevector $\mathbf{k}$

$$X' = \iiint X_{\mathbf{k}}(t) \exp(i\mathbf{k} \cdot \mathbf{x})\, d\mathbf{x}. \tag{4.61}$$

Substituting this form and projecting out the equations for the Fourier components $X_{\mathbf{k}}(t)$ (see Section 4.3.2), we obtain a set of non-linear differential equations. The non-linearities can be modelled, in the usual fashion, by employing a mixing-length formulation for the turbulent diffusivities.

Eichler discusses the energetics and shows that the rate of kinetic energy generation ϵ in the spectral interval $(k, k + \mathrm{d}k)$ by buoyancy force fluctuations is

$$\epsilon(k)\,\mathrm{d}k \propto E_k^{1/2} k^{-3/2} \overline{(\delta\beta/T)}_k\,\mathrm{d}k, \tag{4.62}$$

where E_k is the energy per unit mass in the interval and $\overline{(\delta\beta/T)}_k$ is the buoyancy averaged over a distance π/k.

If the buoyancy forces operate uniformly throughout the system, then the greatest energy release per unit energy input will go into the smallest wavenumbers, i.e. the largest scales, as we would expect. However, if $\overline{(\delta\beta/T)}_k$ has a sharply localized peak then eddies centred on the peak will be favoured.

A rapid variation of the buoyancy force with depth is a feature absent in the models so far discussed, yet it may be a critical factor in deciding whether the fundamental circulation breaks up into smaller eddies. Zeroth-order mixing-length models of the solar convection zone, such as that shown in Fig. 4.1, suggest that the buoyancy force is by no means constant throughout the convection zone. Its variation parallels that of the superadiabatic gradient, which grows with increasing rapidity through the layers where He II, He I and H I successively become ionized. The peak buoyancy is found in a very narrow layer above the level at which hydrogen ionization commences.

It is certainly worth exploring multi-mode numerical simulations with scales corresponding to the total depth of the zone and to the major ionization regions suggested by the mixing-length models. This has yet to be done, but the identification of the supergranulation with the mode originating in the ionization regions will be discussed further in Section 5.4.

If we attribute the origin of the granular-scale convective motions to the highly superadiabatic layer near the upper boundary of the solar convection zone, the granulation is seen to be essentially a 'boundary layer' effect; and in this layer no other motions are of thermodynamic significance (see Sections 2.5.4 and 2.7). There is thus every chance that models of an isolated region with the mean properties of the top of the solar convection zone will provide an adequate representation of the granulation.

In the granulation layer the plasma changes from being optically thick to optically thin. The photon mean free path is now much longer, and radiative smoothing of temperature differences between different parts of the plasma becomes much more effective. This radiative heat exchange must be properly taken into account in the hydrodynamic equations. We are thus led into the realm of 'radiation hydrodynamics', the subject of the next section.

4.4 Radiative heat exchange

4.4.1 Introduction

The source of the Sun's energy is the thermonuclear process in the solar core converting four protons into one alpha particle. The transmutation releases gamma rays which diffuse outwards. In their progress they are absorbed and re-emitted countless times before they emerge at the surface as degraded photons with an effective temperature of 5870 K. Throughout most of the Sun, photon transport is the process which dominates the transfer of energy. It has been estimated that the average time it takes a photon to travel from the centre to the surface is 10^7 years!

The transfer of radiation is governed by the equation expressing conservation of the specific intensity I_λ, the energy per unit wavelength interval crossing unit area normal to the ray direction $\mathbf{\Omega}$, within unit solid angle about $\mathbf{\Omega}$,

$$\partial I_\lambda/\partial t + c(\mathbf{\Omega}\cdot\nabla)I_\lambda = -c\rho k_\lambda I_\lambda + c\rho j_\lambda, \tag{4.63}$$

where k_λ is the mass absorption coefficient and j_λ the mass emission coefficient (Mihalas, 1978). Since the speed of light c is very large compared to convective speeds, we can safely assume that the radiation adopts a stationary state instantaneously. We can then drop the time derivative and consider only the usual equation of transfer

$$(-1/\rho k_\lambda)(\mathbf{\Omega}\cdot\nabla)I_\lambda = I_\lambda - j_\lambda/k_\lambda. \tag{4.64}$$

When the emission and absorption coefficients are known the radiation field may be calculated immediately.

In our discussion of solar hydrodynamics we have been concerned not with the radiation intensity but rather with the radiative energy flux. This is derived from the second of the moments of the intensity, which are generated according to the scheme

$$J = \int I(\Omega)\,\mathrm{d}\Omega/4\pi \tag{4.65}$$

$$H_i = \int \Omega_i I\,\mathrm{d}\Omega/4\pi \tag{4.66}$$

$$K_{ij} = \int \Omega_i\Omega_j I\,\mathrm{d}\Omega/4\pi. \tag{4.67}$$

The first two moments are the mean intensity and the so-called Eddington flux respectively. From Eqn. 4.66 we obtain the radiative energy flux

$$\mathbf{F} = 4\pi\mathbf{H}. \tag{4.68}$$

It would be convenient if we could now proceed to generate closed expressions for the moments J, H_i and K_{ij} by considering only *small* deviations from the state of thermal equilibrium. However, in radiation hydrodynamics this is not possible as we are regularly confronted with substantial deviations from thermal equilibrium. We have thus to solve Eqn. 4.64 in detail.

Defining the opacity, i.e. the volume absorption coefficient, as

$$\kappa_\lambda = \rho k_\lambda, \tag{4.69}$$

and the source function as

$$S_\lambda = j_\lambda / k_\lambda, \tag{4.70}$$

we can write the formal solution of the equation of transfer (4.64) as

$$I_\lambda(\mathbf{x}, \boldsymbol{\Omega}) = \int_C \kappa_\lambda(\boldsymbol{\xi}) S_\lambda(\boldsymbol{\xi}) \exp(-\tau_\lambda)\, \mathrm{d}s, \tag{4.71}$$

where

$$\boldsymbol{\xi} = \mathbf{x} - s\boldsymbol{\Omega}.$$

The path C is taken along the ray in the direction $-\boldsymbol{\Omega}$ through $\mathbf{x}$. The optical thickness τ between the points $\mathbf{x}$ and $\boldsymbol{\xi}$ is defined to be

$$\tau_\lambda(\mathbf{x}, \boldsymbol{\xi}) = \int_C \kappa_\lambda(\boldsymbol{\xi}')\, \mathrm{d}s', \tag{4.72}$$

where

$$\boldsymbol{\xi}' = \mathbf{x} - s'\boldsymbol{\Omega}.$$

The optical thickness has an important physical significance. It is a measure of the number of atomic encounters suffered by the photon along its path. If a photon suffers many absorptions and re-emissions traversing a layer, the layer is said to be optically thick ($\tau > 1$). Otherwise it is optically thin. If the optical thickness between two points is unity, a photon emitted at one point will on average just reach the other before being re-absorbed; as a corollary, the radiation emerging from an atmosphere will arise mainly from points one optical thickness below the surface. Thus

$$I_\lambda(0) \simeq S_\lambda(\tau_\lambda = 1). \tag{4.73}$$

In applying radiative transfer theory to stellar atmospheres optical thickness is measured along the line of sight. However, it is often convenient to express this quantity in terms of the optical *depth* τ_D measured vertically from the surface. Where curvature and horizontal structure may be neglected, i.e. in a plane-parallel atmosphere, we have

$$\tau = \tau_D / \mu, \tag{4.74}$$

where $\mu = \cos\theta$, θ being the angle between the line of sight and the vertical. In terms of optical depth, the intensity of radiation emerging at an angle θ to the normal is given approximately by the Eddington–Barbier relation

$$I_\lambda(0, \mu) \simeq S_\lambda(\tau_D = \mu). \tag{4.75}$$

We will henceforth omit the subscript D when it is clear from the context that optical depth is meant rather than optical thickness.

The geometrical distance corresponding to unit optical thickness can be very large when the gas density is low. Thus the intensity at any given point may depend on the source function and the thermal properties of the atmosphere at quite distant points.

In the analysis of observations, it is important to remember that the Eddington–Barbier relation (4.75) gives only a crude indication of the level whose emission dominates the emergent intensity. It can be seen from Eqn. 4.71 that all heights in the atmosphere for which the quantity

$$C(x_3) = \kappa S\, \mathrm{e}^{-\tau} \tag{4.76}$$

is non-zero must be considered. The quantity $C(x_3)$ is known as the contribution function and measures the contribution of each layer to the emergent intensity. Examples of contribution functions for the photospheric continuum and various lines are shown in Fig. 2.15. They have a FWHM of some 200 km, indicating the degree of non-localization of the emergent intensity. Accordingly, this sets the limit to the height resolution attainable with measurements made at a single wavelength.

More often we are concerned, when analysing the granulation, not with the total emergent intensity but with the fluctuations resulting from a perturbation of the thermodynamic parameters. The manner in which the perturbations at various heights contribute to the observed intensity fluctuations leads to a weighting function that differs from the contribution function (Beckers & Milkey, 1975).

Consider a small change $X'(x_3)$ in an atmospheric parameter that leads to a small change in source function and opacity. When we retain only the linear terms in the Taylor expansions we have

$$S'_\lambda = \frac{\partial S_\lambda}{\partial X} X', \quad \kappa'_\lambda = \frac{\partial \kappa_\lambda}{\partial X} X', \quad \tau'_\lambda = \int_{x_3}^{\infty} \frac{\partial \kappa_\lambda}{\partial X} X'(\xi_3)\, \mathrm{d}\xi_3. \tag{4.77}$$

Substitution in the equation for the emergent intensity normal to the atmosphere leads to the intensity fluctuation

$$I'_\lambda = \int_{-\infty}^{\infty} (\kappa_\lambda + \kappa'_\lambda)(S_\lambda + S'_\lambda) \exp[-(\tau_\lambda + \tau'_\lambda)]\, \mathrm{d}x_3$$

$$- \int_{-\infty}^{\infty} \kappa_\lambda S_\lambda \exp(-\tau_\lambda)\, \mathrm{d}x_3. \tag{4.78}$$

To the first order of small quantities this becomes

$$I'_\lambda = \int_{-\infty}^{\infty} (\kappa'_\lambda S_\lambda + \kappa_\lambda S'_\lambda - \kappa_\lambda S_\lambda \tau'_\lambda) \exp(-\tau_\lambda) \, \mathrm{d}x_3$$

$$= \int_{-\infty}^{\infty} \left(\kappa_\lambda \frac{\partial S_\lambda}{\partial X} + S_\lambda \frac{\partial \kappa_\lambda}{\partial X} \right) \exp(-\tau_\lambda) X'(x_3) \, \mathrm{d}x_3$$

$$- \int_{-\infty}^{\infty} \kappa_\lambda S_\lambda \exp(-\tau_\lambda) \int_{x_3}^{\infty} \frac{\partial \kappa_\lambda}{\partial X} X'(\xi_3) \, \mathrm{d}\xi_3 \, \mathrm{d}x_3, \tag{4.79}$$

or, with a change of order of integration,

$$I'_\lambda = \int_{-\infty}^{\infty} \left[\left(\kappa_\lambda \frac{\partial S_\lambda}{\partial X} + S_\lambda \frac{\partial \kappa_\lambda}{\partial X} \right) \exp(-\tau_\lambda) \right.$$

$$\left. - \frac{\partial \kappa_\lambda}{\partial X} \int_{-\infty}^{x_3} \kappa_\lambda S_\lambda \exp(-\tau_\lambda) \, \mathrm{d}\xi_3 \right] X'(x_3) \, \mathrm{d}x_3$$

$$= \int W_\lambda(x_3) X'(x_3) \, \mathrm{d}x_3. \tag{4.80}$$

The quantity $W_\lambda(x_3)$ is known as the weighting function or line response function and clearly depends on which particular parameter X represents, e.g. temperature or velocity. Such functions are frequently employed in the analysis of observations of the solar granulation (see Section 5.2.2), although the assumption that the opacity fluctuations there are small enough for the linearization to be valid has been questioned by Keil (1980).

However, it is a simple matter to extend all the expansions to second order. The second-order terms are vital when dealing with spatially unresolved observations because the first-order terms then average to zero (see Section 5.5.3).

All weighting functions show roughly the same width as the contribution function though their maxima may be displaced. There is an equal degree of non-localization of the emergent intensity fluctuations.

There is a more subtle aspect of non-localization in radiative transfer that becomes apparent when the content of the source function is examined more closely.

4.4.2 *The source function*

In thermal equilibrium the source function has a simple form, being characterized by a single thermodynamic parameter, the temperature. Moreover there are no intensity gradients, so the source function must be equal to the black-body intensity distribution, the Planck function B_λ. The kinetic energy of the particles is distributed according to Maxwell's law and their internal energy is distributed among the various excitation and ionization states according to the Saha-Boltzmann law. Since the properties of all real systems vary in space and time, true thermal equilibrium can never occur.

Let us examine the case of a simplified atom with just two bound energy levels, E_1 and E_2, of infinite sharpness. Radiative interactions with the nucleus are negligible and energy and momentum are exchanged between the atom and the photon field only via the orbital electrons. The photo-absorption rate causing an upward transition is proportional to the mean intensity $J_{\lambda_{12}}$ at the transition wavelength $\lambda_{12} = hc/(E_2 - E_1)$, the constant of proportionality being the Einstein coefficient B_{12}. The emission rate has a spontaneous component A_{12} and an induced component $B_{21}J_{\lambda_{12}}$, also proportional to the mean intensity. Another important transition process involves the collision of the atom with a free electron. The rate for this process is the product of the collision cross-section C_{12} and the electron density n_e. The continuity equations for the atomic level populations, measured in terms of their number densities n_1, n_2 are then

$$\begin{aligned} Dn_1/Dt &= -n_1 n_e C_{12} - n_1 B_{12} J_{\lambda_{12}} + n_2 A_{21} + n_2 n_e C_{21} + n_2 B_{21} J_{\lambda_{12}} \\ &= -\mathrm{D}n_2/\mathrm{D}t \end{aligned} \tag{4.81}$$

(Mihalas, 1978).

In the solar photosphere the atomic transition times are much shorter than any characteristic time scale of the convective motion, so equilibrium is established instantaneously. In the absence of mass motion, the relative atomic populations are then given by

$$n_2/n_1 = (n_e C_{12} + B_{12} J_{\lambda_{12}})/(A_{21} + B_{21} J_{\lambda_{12}} + n_e C_{21}), \tag{4.82}$$

which is the equation of statistical equilibrium.

The Einstein coefficients are properties of the atom and are related to one another as follows:

$$B_{12} = B_{21} \varpi_2 / \varpi_1 \tag{4.83}$$

$$A_{21} = B_{21} 2hc/\lambda_{12}^3, \tag{4.84}$$

where ϖ_1, ϖ_2 are the statistical weights of the two levels. The collision cross-sections depend on temperature and can be evaluated using the Maxwellian velocity distribution; then

$$C_{12} = C_{21}(\varpi_2/\varpi_1) \exp(-hc/\lambda_{12} kT). \tag{4.85}$$

Substitution in (4.82) yields

$$\frac{n_2\varpi_1}{n_1\varpi_2}=\frac{B_{21}J_{\lambda_{12}}+n_e C_{21}\exp(-hc/\lambda_{12}kT)}{B_{21}J_{\lambda_{12}}+A_{21}+n_e C_{21}}. \quad (4.86)$$

If we assume isotropy and treat stimulated emission as negative absorption, we can evaluate the source function for the line transition, obtaining

$$S_{\lambda_{12}}=\frac{j_{\lambda_{12}}}{k_{\lambda_{12}}}=\frac{n_2 A_{21}hc/4\pi\lambda_{12}}{n_1 B_{12}hc/4\pi\lambda_{12}-n_2 B_{21}hc/4\pi\lambda_{12}}. \quad (4.87)$$

Using Eqn. 4.86 this can be simplified to

$$S_{\lambda_{12}}=(J_{\lambda_{12}}+\epsilon' B_{\lambda_{12}})/(1+\epsilon'), \quad (4.88)$$

where

$$\epsilon'=[1-\exp(-hc/\lambda_{12}kT)]\,n_e C_{21}/A_{21}. \quad (4.89)$$

The source function thus contains two terms. The first is a scattering term representing photon absorption followed immediately by re-emission over all angles. (It should be noted that this is the result of successive single photon events and is not scattering in the quantum-mechanical sense, in which the transition is virtual.) The second is a thermal term which arises when collisional de-excitation follows the absorption of a photon. Only this second process transfers photon energy and momentum to the free electron gas. The free electrons equipartition the kinetic energy both among themselves (hence the assumption of a Maxwellian distribution above) and with the atoms through elastic collisions; the first process is much more rapid than the second.

Inelastic collisions, producing a transition which is followed by radiative de-excitation, couple the electron energy reservoir to the photon field. The electron-photon coupling is measured by the parameter ϵ', which is basically a measure of the ratio of the electron collision rate to the spontaneous decay rate. In stellar atmospheres, the relatively low electron densities ensure that $\epsilon' < 1$. Under these circumstances, the destruction or thermalization length – the distance travelled by a photon before it is destroyed rather than scattered – becomes greater than the scale height of the atmosphere. The source function itself is thus non-locally determined.

In the *low* solar photosphere, the situation is not so extreme. There, collisions are frequent enough for the atomic level populations (Eqn. 4.86) to retain the equilibrium values appropriate to the local kinetic temperature. This is known as *local thermodynamic equilibrium* (LTE). It implies that, as in true thermal equilibrium, the source function is equal to the local Planck function, i.e.

$$S_\lambda = B_\lambda, \quad (4.90)$$

although the intensities in the inward and outward directions may separately depart considerably from the Planckian value.

The simplified case considered above is easily generalized to multi-level atoms and to ionization (Mihalas, 1978); the equations become more cumbersome but the basic distinction between scattering and collisional processes remains. The relative infrequency of collisions will always produce non-LTE effects in stellar atmospheres. The source function may depart significantly from the Planck function, and the atomic level populations may depart widely from those calculated from the Saha-Boltzmann equation.

For example, Athay & Lites (1972) have analysed the equilibrium of the iron atom in the solar atmosphere and have concluded that the ionization equilibrium departs from the Saha (LTE) value above $\tau = 0.05$ and that the excitation equilibrium, for low-lying atomic levels, departs from the Boltzmann (LTE) values above $\tau = 10^{-3}$-10^{-4}. Non-LTE effects need to be taken into account in analysing the profiles of most solar lines.

The general case requires the simultaneous numerical solution of Eqn. 4.71 along chosen rays together with the set of statistical equilibrium equations. Even in plane-parallel (one-dimensional) atmospheres, the problem is complex and the reader is referred to Mihalas (1978). In multi-dimensional problems, resort is often made to the simplified versions of the transfer equation discussed below.

4.4.3 *The Eddington and diffusion approximations*

In granulation models the treatment of radiative transfer has often been grossly simplified by invoking certain approximations. In this section we shall derive these approximations, but shall postpone a discussion of their validity to the following section.

If we drop the wavelength subscripts temporarily, the equation of radiative transfer may be written

$$(\mathbf{\Omega} \cdot \nabla) I = \kappa (S - I). \tag{4.91}$$

If we multiply this equation by the direction cosine Ω_i, integrate each term over solid angle, and recall the definitions of the moments (Eqns. 4.65-7), we obtain a moment equation

$$\nabla \cdot \mathbf{H} = \kappa (S - J). \tag{4.92}$$

Multiplying by $\Omega_i \Omega_j$ and integrating produces another moment equation

$$\partial K_{ij} / \partial x_j = -\kappa H_i, \tag{4.93}$$

and so on.

This system of moment equations cannot be truncated and closed except under special circumstances. Close to thermal equilibrium it can be assumed that the radiation field is almost isotropic, and the general expansion of the radiation field in spherical harmonics

$$I(\mathbf{x}, \Omega) = \sum_{n=0}^{\infty} \sum_{m=-\infty}^{\infty} I_{mn}(\mathbf{x}) Y_{mn}(\theta, \phi), \tag{4.94}$$

can be restricted to the zeroth- and first-order terms. Substituting in the definitions of the moments (Eqns. 4.65-7) we then find that

$$K_{ij} = J\delta_{ij}/3. \tag{4.95}$$

This relation provides the closure necessary to truncate the system of moment equations. The quantities K and H can now be eliminated from Eqns. 4.92 and 4.93 to yield the following equation for the mean intensity

$$\frac{1}{\kappa} \nabla \cdot \left(\frac{1}{\kappa} \nabla J \right) = 3(J - S). \tag{4.96}$$

This is termed the 'general' Eddington approximation and was first derived by R. G. Giovanelli (1959).

The divergence of the heat flux may be evaluated directly:

$$\nabla \cdot \mathbf{F} = 4\pi \nabla \cdot \mathbf{H} = -4\pi\kappa(J - S). \tag{4.97}$$

The computational simplification provided by the approximation is apparent when one counts the number of independent parameters describing the radiation field at any point. In one-dimensional atmospheres there are two non-zero coefficients in the expansion (4.94), which can be evaluated from just two intensities, an ingoing and an outgoing intensity along a single ray. In two dimensions there are three parameters, which require two independent rays. In three dimensions four parameters are required, again provided by only two rays.

In LTE, the Planck function may be substituted for the source function.

At great depths, the opacity will be very large and the variation of the source function over the regions contributing to the intensity integral (4.71) will be small. It is then sufficient to consider only the linear variation of the source function in the neighbourhood of $\mathbf{x}$,

$$\begin{aligned} S(\xi) &= B(\mathbf{x}) + (\nabla B) \cdot (\xi - \mathbf{x}) \\ \kappa(\xi) &= \kappa(\mathbf{x}). \end{aligned} \tag{4.98}$$

In this case, the integral (4.71) can be evaluated exactly giving

$$I(\mathbf{x}, \Omega) = B(\mathbf{x}) - \nabla B(\mathbf{x}) \cdot \Omega/\kappa. \tag{4.99}$$

As in the Eddington approximation, the intensity field is almost isotropic; in addition, we find

$$J = B, \tag{4.100}$$

and

$$\mathbf{H} = -\nabla B/3\kappa. \tag{4.101}$$

Eqn. 4.101 represents the diffusion approximation and is valid when the photon mean free path is short compared to the scale height. It is more restrictive than the Eddington approximation.

Restoring the suppressed wavelength suffixes and integrating over wavelength yields the total heat flux

$$\begin{aligned} \mathbf{F} = 4\pi \int \mathbf{H}_\lambda \, \mathrm{d}\lambda &= -\frac{4\pi}{3} \nabla T \int \frac{1}{\kappa_\lambda} \frac{\mathrm{d}B_\lambda}{\mathrm{d}T} \mathrm{d}\lambda \\ &= -\frac{4\pi}{3\kappa_\mathrm{R}} \frac{\mathrm{d}B}{\mathrm{d}T} \nabla T, \end{aligned} \tag{4.102}$$

where κ_R is the Rosseland mean opacity, defined by

$$1/\kappa_\mathrm{R} = \left[\int (1/\kappa_\lambda)(\mathrm{d}B_\lambda/\mathrm{d}T)\,\mathrm{d}\lambda\right] / (\mathrm{d}B/\mathrm{d}T). \tag{4.103}$$

Using

$$\mathrm{d}B/\mathrm{d}T = \mathrm{d}\left(\int B_\lambda \,\mathrm{d}\lambda\right)/\mathrm{d}T = \mathrm{d}(\sigma T^4/\pi)/\mathrm{d}T = 4\sigma T^3/\pi, \tag{4.104}$$

where σ is the Stefan-Boltzmann constant, we obtain

$$\mathbf{F} = -16\sigma T^3 \nabla T/3\kappa_\mathrm{R}. \tag{4.105}$$

This is a conduction equation, the coefficient of radiative conductivity being

$$K = 16\sigma T^3/(3\kappa_\mathrm{R}). \tag{4.106}$$

It will be clear from the derivation that the diffusion approximation is a perfectly adequate and self-consistent description of radiative energy transfer within the subphotospheric layers – e.g. throughout most of the solar convection zone. On the other hand, the diffusion approximation is not valid in stellar atmospheres.

The self-consistency of the Eddington approximation is not immediately obvious. In optically thick layers it is equivalent to the diffusion approximation, since it is based on the same degree of isotropy. The extent to which the assumed isotropy breaks down in optically thin layers is discussed in the next section.

4.4.4 *Radiative smoothing*

The transfer of energy by radiative processes is always accompanied by momentum transfer, but the resultant viscous effects are negligibly small. Hence in the internal energy equation (3.4) we need only include the radiative flux term from Eqn. 4.97,

$$\nabla \cdot \mathbf{F} = 4\pi \int \kappa_\lambda (S_\lambda - J_\lambda)\,\mathrm{d}\lambda. \tag{4.107}$$

In order to gain some insight into the role of this term let us first consider a stationary medium and ignore all other means of heat transfer. Then the energy equation becomes

$$\rho C_p\,\partial T/\partial t = -\,4\pi \int \kappa_\lambda (S_\lambda - J_\lambda)\,\mathrm{d}\lambda. \tag{4.108}$$

If we assume the opacity to be wavelength-independent (the so-called grey approximation) and LTE to obtain, this simplifies to

$$\rho C_p\,\partial T/\partial t = -\,4\pi\kappa(\sigma T^4/\pi - J), \tag{4.109}$$

where the mean intensity is derived from the equation of transfer (4.71)

$$J(\mathbf{x}) = \iiint \kappa\,\frac{\sigma T^4}{\pi}\,\exp(-\tau)\,\frac{\mathrm{d}\boldsymbol{\xi}}{4\pi\,|\boldsymbol{\xi} - \mathbf{x}|^2}, \tag{4.110}$$

the integral being taken over all space.

Eqns. 4.109 and 4.110 are difficult to solve in the general case. However, Spiegel (1957) has analysed the simplified case of a uniform medium of temperature T_0 and opacity κ_0 containing small temperature perturbations with wavenumber k. The linearized equations then admit solutions of the form

$$T' \propto \exp(nt)\exp(\mathrm{i}\mathbf{k}\cdot\mathbf{x}). \tag{4.111}$$

The growth rate n is the only eigenvalue of the system and is always negative, i.e. the temperature perturbation decays at a rate

$$-n(k) = [1 - (\kappa_0/k)\cot^{-1}(\kappa_0/k)]/t_c \tag{4.112}$$

where

$$t_c = \rho_0 C_p / 16\kappa_0 \sigma T_0^3. \tag{4.113}$$

The factor t_c is the local cooling time. It gives the cooling rate when the perturbation is optically thin ($\kappa_0/k \ll 1$) - the case when photons are exchanged directly between the hot and cool regions without intervening absorption. As the optical thickness of the perturbation increases, the photon exchange is accomplished with greater difficulty and the cooling rate decreases.

In stratified atmospheres the eigenvalues are no longer single-valued but form a continuous spectrum. Furthermore, some of the eigenvectors are localized, in that they are significant only over a narrow height range; the associated cooling rates are given to a good approximation by an expression due to Ulrich (1970c)

$$-n = (1 - \tau_m \cot^{-1}\tau_m)/t_c, \tag{4.114}$$

where τ_m is a mean optical thickness defined as

$$\tau_m = [(k/\kappa_0)^2 + (1/\tau)^2]^{-1/2}. \tag{4.115}$$

τ is the optical distance from the 'top' of the atmosphere.

On the other hand, the non-local eigenvectors reflect the continual transfer of radiative energy from deep to higher layers. This is controlled by the deep-seated heat supply and therefore the associated cooling times are longer. A temperature perturbation generally distributed throughout the solar atmosphere will first decline rapidly in the outer layers due to local cooling, but the long-term behaviour will be a uniform slow decrease as the perturbation in the optically thick layers leaks away through the atmosphere. Expressions (4.112) and (4.113) provide lower and upper bounds for the cooling times in the low photosphere.

The cooling times for (*a*) an optically thin perturbation and (*b*) one of granular scale ($k_\perp = 2\pi/2.5\ \text{Mm}^{-1}$) are given in Table 4.2, after Lévy (1974). Both are short compared to the convective time scale. Radiative exchange thus rapidly ensures the vanishing of the radiative heat-flux gradient, so that

$$\nabla \cdot \mathbf{F} = 4\pi \int \kappa_\lambda (S_\lambda - J_\lambda)\, d\lambda = 0. \tag{4.116}$$

Table 4.2. *Radiative cooling times in the solar atmosphere for an optically thin perturbation (t_c) and for a perturbation of horizontal wavenumber* $k_\perp = 2\pi/2.5\ \text{Mm}^{-1}$ $(1/|n|)$

Height (km)	t_c (s)	$1/\|n\|$ (s)
−50	0.015	910
0	0.28	230
50	1.9	63
100	7.6	28
150	16	30
200	25	36
300	53	48
400	83	83
500	110	110

This condition is known as *radiative equilibrium*. As we approach the base of the atmosphere the cooling time increases owing to the increasing optical thickness (see Eqns. 4.114 and 4.115) and Eqn. 4.116 ceases to be valid. At these depths convection gives rise to a fluctuating heat flux (see Section 5.3).

The structure of a grey radiative equilibrium atmosphere subject to a varying flux at its base has been investigated by Kneer & Heasley (1979). They treated a simplified case in which the opacity shows no horizontal variation but depends on depth according to the relation

$$\tilde{\kappa} = \exp(-\tilde{x}_3); \tag{4.117}$$

$\tilde{x}_3$ is measured in units of the opacity scale height. Fig. 4.5 shows the amplitude of the mean intensity (or source function) fluctuation as a function of optical depth for intensity perturbations of various horizontal wavelengths (also measured in units of the opacity scale height) at the bottom of the atmosphere.

The full lines show the exact solution, the dashed lines the Eddington approximation. Two features are readily apparent:

(1) the Eddington approximation becomes steadily worse as the wavelength of the horizontal perturbation becomes shorter, i.e. the atmosphere becomes effectively more two-dimensional. The limited anisotropy allowed by the Eddington closure is clearly inadequate;

(2) decreasing the wavelength allows the source function and hence intensity fluctuations to be smoothed out more and more by horizontal transfer. Since the mean intensity and source function are equal in radiative equilibrium (Eqn. 4.116), Eqn. 4.88 shows that both quantities must be equal to the thermal radiation field B whether or not LTE obtains. It follows that the effect of horizontal radiative transfer is to smooth out the temperature fluctuations as well as the intensity fluctuations, a process which may be described as thermal smoothing.

At the upper boundary ($\tau \sim 10^{-2}$) the intensity fluctuations are reduced by two orders of magnitude when the horizontal wavelength of the perturbation is still ten times the scale height in the photosphere, corresponding to 600–1200 km on the Sun. Hence any perturbation induced by a *convective* heat input will likewise be effectively damped out by horizontal radiative transfer. This must be accurately treated, but we cannot use the Eddington approximation since, as we have seen, it is in error by an order of magnitude in the upper atmosphere.

Kneer & Heasley's analysis is grossly oversimplified in that no account is taken of horizontal opacity fluctuations induced by the perturbations. Their model predicts that the temperature fluctuations decay with a scale height equal to about twice the opacity scale height, i.e. about 200 km in the photosphere, whereas in Chapter 5 we shall see that the temperature perturbations in the solar atmosphere appear to decay more steeply. This can probably be attributed to an additional effect known as 'photon channelling'. If the opacity decreases in cool regions, photons escape more easily from them than from a neighbouring high-opacity hot region. Photons from the latter seek to escape through the cool 'channel', thus reducing the mean intensity of the hot region and increasing that of the cool. Radiative smoothing is thus more effective than a constant-opacity model would indicate (Cannon, 1970).

Fig. 4.5. Amplitude of source function fluctuation versus optical depth in an atmosphere in radiative equilibrium. The curves are labelled by the horizontal wavelength of the flux perturbation in units of the opacity scale height. Solid lines are the exact solution, dashed and dotted lines the Eddington approximation solutions for two different upper boundary conditions (Kneer & Heasley, 1979).

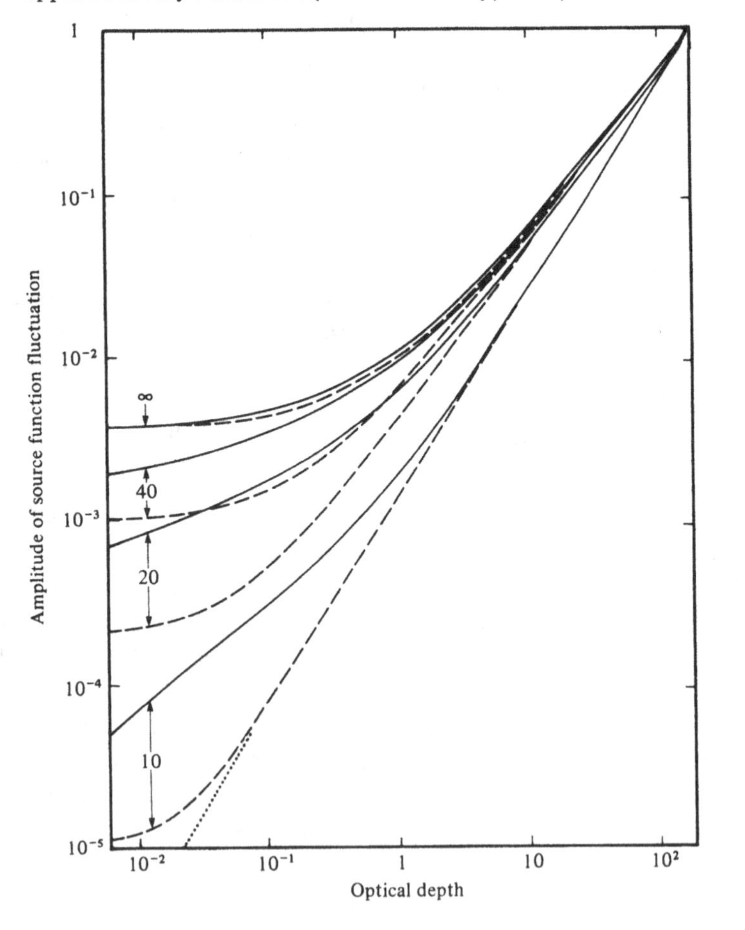

The real solar atmosphere is neither grey nor is it exactly in radiative equilibrium. Despite radiative smoothing, temperature fluctuations will persist to some extent and will influence the line radiation for which LTE does not apply. The question thus arises of the role played by multi-dimensional radiative transfer in deducing the fluctuations of thermodynamic properties from line observations.

This problem has been addressed by Kneer (1980). Instead of imposing radiative equilibrium, he introduces a thermal perturbation with a sinusoidal horizontal variation whose amplitude is independent of height. The fluctuation of the mean intensity as a function of optical depth is shown in Fig. 4.6 for a line with a thermal coupling parameter $\epsilon' \sim 10^{-4}$ (see Eqns. 4.88 and 4.89). This

Fig. 4.6. Amplitude of the mean intensity fluctuation J' relative to the amplitude of the thermal fluctuation B' as a function of optical depth in a line formed by coherent scattering ($\epsilon' = 10^{-4}$). The curves are again labelled by the horizontal wavelength of the thermal perturbation in units of the opacity scale height (Kneer, 1980).

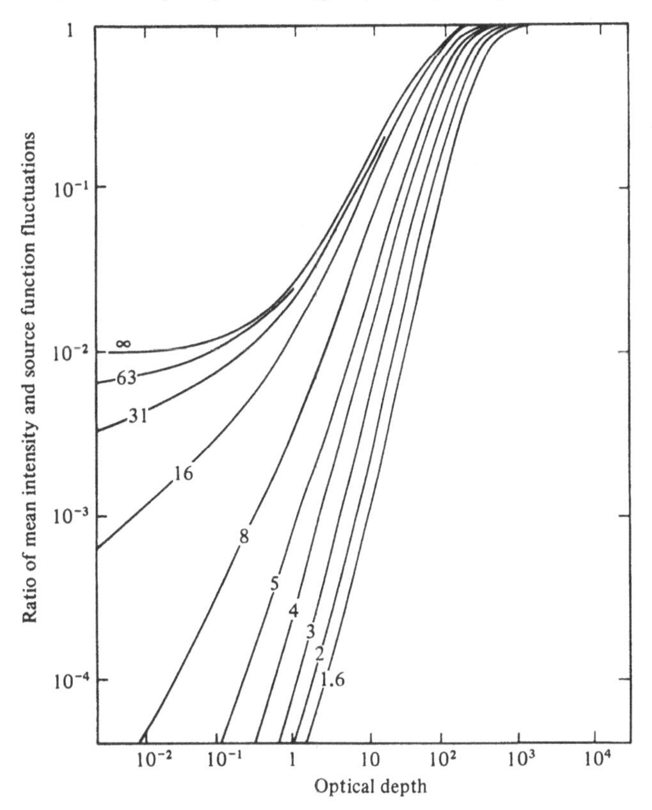

is also the fluctuation of the scattering term in the source function since the thermal contribution is negligible. The curve $\Lambda_\perp = \infty$ represents the limiting case of an infinitely long horizontal wavelength and thus shows the result of ignoring horizontal transfer and treating each column of the atmosphere as an independent plane-parallel slab. Evidently radiative transfer in three dimensions dampens the fluctuation of the scattering term; the smoothing increases as ϵ' decreases, i.e. as the degree of departure from LTE increases.

The choice of $\epsilon' \sim 10^{-4}$ is a lower extreme for photospheric lines, and accordingly the smoothing estimated in this case is an upper bound.

At $\Lambda_\perp \sim 2000$–3000 km in the solar photosphere the total source function fluctuation shows negligible smoothing, while at $\Lambda_\perp \sim 300$–500 km it is an order of magnitude even at the level of the continuum, $\tau = 1$. Treating the atmosphere as a set of independent columns thus underestimates the thermal fluctuations by more than an order of magnitude. We may thus conclude that observational results which make no allowance for multi-dimensional radiative transfer effects are reliable at typical granular scales but, at intergranular scales, may underestimate the amplitude of the temperature fluctuations.

Fortunately, smoothing has less effect on velocity determinations. A Doppler shift is produced by scattering (i.e. absorption followed by re-emission) and does not involve an interaction between the photon and the free electron gas. It thus provides a measure of the mass motion at optical depth unity along the line of sight (Gouttebroze & Leibacher, 1980).

The subject of non-uniform radiative transfer is rapidly developing; for different points of view the interested reader is referred to Cannon (1976), Jones & Skumanich (1980) and Owocki & Auer (1980).

Additional notes

The properties of hypothetical giant cells in the solar convection zone have been investigated by Latour, Toomre & Zahn (1983, *Solar Physics*, **82**, 387–400) by means of a single-mode anelastic model. Integrating the full time-dependent equations for a polytropic atmosphere, K. L. Chan, S. Sofia & C. L. Wolff (1982, *Astrophysical Journal*, **263**, 935–43) find a range of vertical cells corresponding to the varying pressure scale height (Section 4.3.4).

A. Legait (1982, *Astronomy and Astrophysics*, **108**, 287–95) has discussed the dynamical consequences of the Eddington and diffusion approximations and has formulated a procedure for interpolating between them (Section 4.4.3).

5

INTERPRETATION OF THE GRANULATION AND SUPERGRANULATION

5.1 Introduction

In the previous two chapters we have tried to give an account of the theoretical work on convection (together with the less extensive laboratory experiments) which has some degree of relevance to the problem of the interpretation of the granulation. We have not, hitherto, discussed the theory of convection in relation to the observed features of the granulation, which are exhaustively described in Chapter 2. This we now propose to do.

The comparison between theory and observation is effected by the construction of models - models, in this case, of the inhomogeneous structure of the solar photosphere and upper layers of the convection zone. An inhomogeneous model is one that takes account of the horizontal as well as vertical variations in the various hydrodynamic and thermodynamic quantities. The various attempts to derive such models are described in Section 5.2. They range from the totally empirical to the (almost) completely self-consistent. Each approach has its fundamental limitations, whether due to uncertainty in the data or in the numerical simplification imposed; nevertheless a consistent picture of the granulation emerges. This is a most important conclusion: the derivation of an adequate inhomogeneous model of the Sun is almost the sole astrophysical test of the basic theory of convection in highly stratified fluids.

In Section 5.3 we shall confront the features common to these models with the basic observational facts. The topics discussed include the mean cell size, lifetime, granulation near the extreme limb, height of overshoot, convective heat flux, direction of cellular motion and turbulence spectrum.

Our understanding of the supergranulation is insufficient to allow much meaningful comparison with observations. Our knowledge is almost exclusively based on observation, the interpretation of which is the concern of Section 5.4.

Finally, we may ask whether our knowledge of the solar granulation is such that we can undertake an investigation of convection in other stars. Here spatially resolved observations are not available, only observations in light integrated over the stellar disk. Accordingly, in Section 5.5 we discuss the convective effects apparent in solar line profiles observed without spatial resolution, in particular, their relative displacements and asymmetries. In the Sun these effects are small but are specific to a convective heat flow.

We conclude in Section 5.6 with a brief survey of the prospects of a direct investigation of stellar granulation based on such diagnostics.

A list of the principal symbols employed in this chapter will be found in Table 3.1.

5.2 Models of the solar granulation

5.2.1 Introduction

Before proceeding to discuss individual models, it is important to clearly state the main reasons for regarding the granulation as a convective phenomenon. These reasons are threefold, the first two being based on observational evidence and the third being inferred from our zeroth-order knowledge of the physical conditions in the solar convection zone:

(1) the granulation shows a clearly defined cellular pattern with a relatively narrow distribution of cell sizes (Sections 2.3.1 and 2.3.3). Buoyancy-driven convection is the only physical mechanism known to produce such a pattern, apart from surface tension and other forces known to be absent;

(2) there is a strong correlation in both sense and magnitude between brightness and velocity in the photosphere at granular scales (Sections 2.4.2 and 2.5.5). This is qualitatively what one would expect for buoyancy-driven convection, since the hot gases must move upward;

(3) if the granulation is to be identified as a convective phenomenon, then we must find a source of energy for driving the convective motions. It is now generally agreed that this source is the liberation of hydrogen ionization energy by upward-moving material. As can be seen in Fig. 4.1, the fraction of hydrogen ionized increases from 10% to 90% between depths of 300 km and 7000 km in the convection zone. Throughout this range, the ionization energy is a substantial fraction of the total energy. At a depth of 300 km a dramatic decrease occurs as a result of the almost complete recombination of the hydrogen ions. If we imagine an elementary volume of gas moving upwards through such a region of decreasing ionization, then it is clear that as it moves upwards it liberates ionization energy. Thus the buoyancy of the element is increased, and it continues its upward journey, provided the forces opposing its motion are sufficiently small. In other words, the

medium is convectively unstable. This instability is very high when the degree of ionization changes rapidly with depth, which in the photosphere occurs over the first few hundred kilometres.

These are the main reasons, therefore, which lead us to identify the photospheric granulation with convection, the driving force owing its origin primarily to the release of hydrogen ionization energy several hundred kilometres below the photosphere.

Spiegel (1964) proposed an alternative explanation. He suggested that rather than being thermally generated, the granulation may be mechanically driven by the supergranulation; however, no detailed theory of such a process has yet been worked out. Indeed, the increasing respectability of purely convective models now renders the search for other explanations unnecessary.

Whilst there is no reason to doubt that the liberation of hydrogen ionization energy is the principal mechanism driving the convective motions, another glance at Fig. 4.1 reveals a secondary consideration in just those layers where we expect the granulation to occur. The buoyancy of the upward moving element, as measured by the superadiabatic gradient, is greatest not within the hydrogen ionization zone but above it, less than 100 km below the base of the photosphere. Prior to ionization the hydrogen opacity increases with increasing temperature and this alone reduces the effectiveness of radiative heat transfer in the layers immediately below the solar surface. The temperature gradient thus steepens and makes these layers convectively unstable. Even in the convective state the temperature gradient remains steep because the convective transport is inefficient in this region (see Section 4.2.2). The significance of this region as a source of granular motions was first remarked by Plaskett (1955) and later reiterated by Cloutman (1979a). It will be discussed further in Section 5.2.4.

Among other shortcomings, the zeroth-order model of the solar convection zone described in Section 4.2 (and shown in Figs. 4.1 and 5.1) departs from reality in one very important respect: it gives no expression to horizontal variations in such physical quantities as the temperature, density, pressure, absorption coefficient, etc., although the very existence of the photospheric granulation implies that such variations must be present. Obviously, there are two basic reasons for developing an inhomogeneous model of the solar photosphere and the convection zone as a whole:

(1) only an inhomogeneous model of the photosphere can provide a proper theoretical basis for explaining, on the one hand, the observed values of such quantities as the granular contrast and its centre-to-limb variation and, on the other hand, the asymmetry in the profiles of Fraunhofer lines and the centre-to-limb variation of their wavelengths;

(2) only an inhomogeneous model of the convection zone, together with an appropriate radiative transfer theory for non-uniform media, can give an exact form for the radiative term or terms which, as we have seen in Section 4.4, have to be included in the basic hydrodynamic equations governing astrophysical convection.

5.2.2 *Empirical inhomogeneous photospheric models*

Inhomogeneous models of the solar photosphere are almost always obtained by perturbing a model of the mean atmosphere. The mean model is based on observations of the wavelength dependence and centre-to-limb variation of the emergent intensity. The temperature stratification of the model is then adjusted until the observed quantities are, more or less, reproduced. Since the mean model is assumed to be stationary, hydrostatic equilibrium is invoked to provide the run of pressure. Models which are based on both observation and theory are known as semi-empirical. Examples are provided by the Bilderberg

Fig. 5.1. Semi-empirical mean models of the solar photosphere and upper convection zone. Temperature, pressure (in units of $10^2\,\mathrm{N\,m^{-2}}$) and density (in units of $10^{-6}\,\mathrm{kg\,m^{-3}}$) are shown as functions of height and optical depth for the mixing-length model of Spruit (1974) (broken line) and the two-dimensional model of Nelson (1978) (solid line).

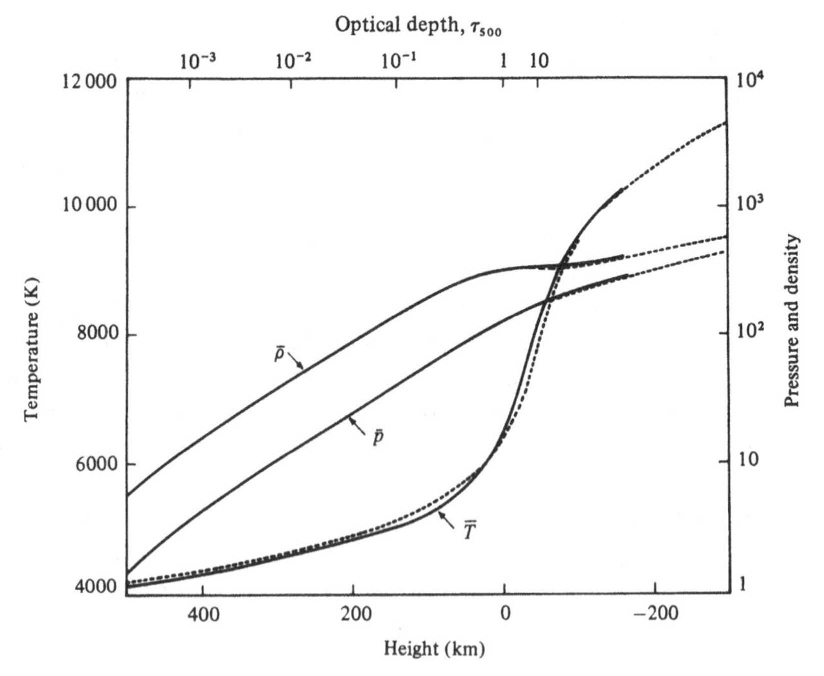

Continuum Atmosphere (BCA) (Gingerich & de Jager, 1968) and the Harvard-Smithsonian Reference Atmosphere (HSRA) (Gingerich, Noyes, Kalkofen & Cuny, 1971).

In reading the rest of the chapter one should bear in mind that the mean atmosphere becomes convectively unstable below the level $\tau = 1$ in the HSRA and below 10 km above $\tau = 1$ in the BCA. Continuum radiation thus tends to originate somewhat below the level of convective instability, line radiation above it (see Fig. 2.15). Thus granulation observations in Fraunhofer lines refer almost exclusively to the so-called interfacial region where overshooting occurs.

The development of inhomogeneous models of the photosphere began with Albrecht Unsöld, who was the first to suggest that the solar atmosphere could be treated as a set of columns in horizontal pressure balance with one another but each having independently specified distributions of temperature and vertical velocity (see Böhm, 1954). The distributions are chosen so as to reproduce measurements of various solar quantities such as granular contrast and convective velocity, as well as the centre-to-limb variation of the strengths and profiles of photospheric Fraunhofer lines. The number of free parameters is so large that no unique model can result from such a procedure. Nevertheless, a series of such models - de Jager (1954, 1959), Böhm (1954), Voigt (1956, 1959), Schröter (1957) - all display the same gross characteristics despite differences in detail.

The culmination of the series was the Utrecht Reference Atmosphere (Heintze, Hubenet & de Jager, 1964), whose photospheric portion was based on Voigt's 3-stream model of 1956. It possessed a stationary component with the properties of the mean atmosphere, a hot column with an upflow and a cool column with a downflow. The differences between the columns and the mean atmosphere were taken to vanish above 100 km ($\tau = 0.1$), but to increase with depth to values of $+350/-400$ K in temperature and $+2.3/-1.7$ km s^{-1} in vertical velocity at the (same) geometric height corresponding to $\tau = 1$ in the stationary column. At equal *optical* depths $\tau = 1$ in the hot and cold columns the temperature differences are $+100/-230$ K. Despite the degree of simplification imposed on this model and the authors' caution in not investing it with spurious respectability, these basic properties are reproduced in all subsequent analyses.

One of the more easily measured properties of the solar granulation is the rms brightness fluctuation at the centre of the disk (Sections 2.5.2 and 2.5.3). This is usually interpreted as being due solely to a temperature fluctuation small enough for the variation to be treated as a linear perturbation (see Section 4.4.1). Keil & Canfield (1978) and Kneer, Mattig, Nesis & Werner (1980) thereby derived the height variation of the rms temperature fluctuations shown in Fig. 5.2(*a*). The run of temperature fluctuation shows two noteworthy features, a very steep gradient in the lowest layers of the photosphere and a much shallower gradient in the upper photosphere.

Kneer *et al.* stressed the smallness of the correlation between the temperature fluctuations in the lower photosphere and those in the rest of the atmosphere. Durrant & Nesis (1981) have expressed doubts as to whether the temperature field in the upper region should be treated as a granular phenomenon at all.

When scaled to a common rms brightness fluctuation of 12%, all analyses require an rms temperature fluctuation of about 400–500 K at $x_3 = 0$, dropping to less than 100 K within less than 50 km above this level.

Fig. 5.2. Empirical determinations of the height dependence of the granular temperature fluctuations and vertical velocities in the solar atmosphere.
(a) Rms values. The temperature fluctuations are adjusted to give a standard continuum brightness fluctuation of 12% around 500–550 nm. The vertical velocities are taken from Fig. 2.24.
(b) Amplitude (half peak-to-peak). The band of temperature fluctuations is taken from Altrock & Musman (1976), the run of vertical velocity from Keil (1980c).

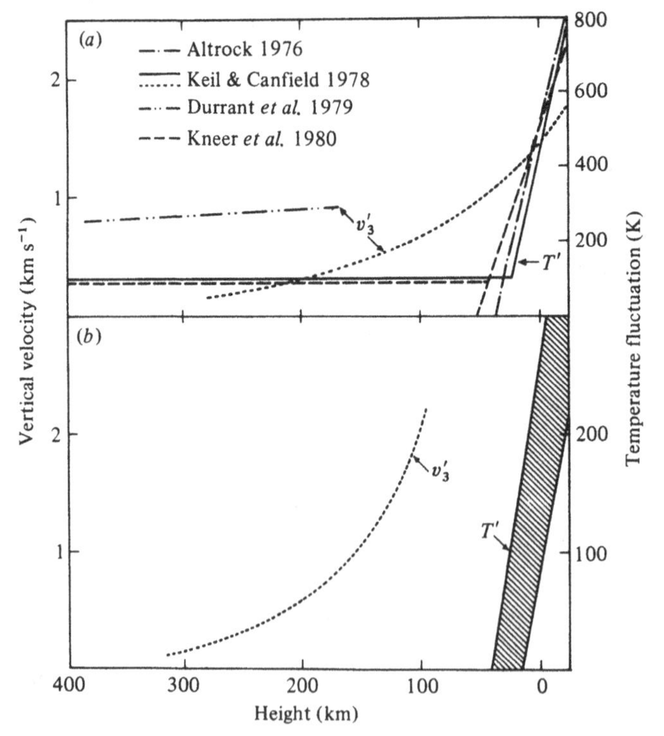

Altrock & Musman (1976) have attempted to measure the height variation of the temperature perturbations in individual granules, but their sample was very small (three granules and three intergranular regions). The general run of temperature difference between the granules and the mean, shown in Fig. 5.2(*b*), closely mirrors that derived from rms measurements and shows no component in the upper photosphere. On the other hand, they found a slight temperature *excess* in the intergranular region above 150 km, but the intergranular results might well have been influenced by seeing since the mean contrast of their sample was only 9.5% at λ500 nm. The analysis of the velocity fluctuations remains unsatisfactory. The conflicting determinations of the height dependence of the rms vertical velocity of the granulation due to Keil & Canfield (1978) and to Durrant *et al.* (1979), described in Section 2.5.4, are shown again in Fig. 5.2(*a*). These values were obtained from a linear analysis using velocity weighting functions (see Section 4.4.1), but Keil (1980a) has since shown that this approach is not valid. Neither the temperature correlations nor the non-linear effects of the velocities can be neglected; their inclusion leads to a steeper velocity gradient in the atmosphere.

Keil (1980b, c) took account of all these effects in his analysis of a time sequence of observations from which the oscillatory field had been removed by filtering. As we saw in Section 2.4.2, he obtained a picture, averaged over many examples, of the instantaneous vertical velocities in granules and intergranular lanes at various heights. The distribution of velocity amplitude is shown in Fig. 5.2(*b*). Whilst the absolute values are uncertain due to uncertainties in the correction procedure, the gradient of the velocity amplitude is more reliable and yields a scale height of about 80 km.

It seems safe to conclude that the granular temperature fluctuations effectively vanish at a height of 50 km above the level of continuum formation ($\tau = 1$) whilst the granular vertical velocity field overshoots, with ever-decreasing strength, to levels above 300 km. At 500 km the granular velocities are no longer detectable and are replaced by small-scale disturbances which vary on a much shorter time scale than the granulation (Keil, 1980b).

It should be noted that the ranges of physical heights for which the separate determinations of *granular* temperature and velocity fluctuations are trustworthy do not overlap. The vertical velocities in the continuum forming layers can be obtained only by extrapolation. The use of an exponential curve to describe the height variation then leads to implausibly large, indeed supersonic, values at $\tau = 1$!

Turning now to the need for multi-dimensional radiative transfer theory in realistically modelling the photosphere, we should first recall the discussion in Section 4.4.4. There it was shown that, for horizontal scales typical of the granulation, the fluctuations in thermodynamic quantities may be determined reliably from observations by treating each column of the atmosphere as a section of a plane-parallel model – an assumption which underlies the analyses described above. An immediate corollary is that each line of sight may be treated independently, since the emergent intensity depends solely on the distribution of temperature and pressure along that ray.

Wilson (1962 and subsequently) was the first to incorporate the horizontal variation of atmospheric quantities directly into solutions of the transfer equation. In his 1964 paper, he considered continuum radiation (which is not influenced by velocity fields) and prescribed a two-dimensional temperature fluctuation that gave rise to a source function fluctuation of the form

$$B'(x_1, x_3) = (A + Bx_3)\exp(-px_3)\cos k_1 x_1. \tag{5.1}$$

The horizontal wavenumber $k_\perp \equiv k_1$ was chosen to represent a mean granular scale, viz. $k_\perp = 2\pi/1400\,\mathrm{km}^{-1}$. Wilson's original concern was to reproduce the centre-to-limb measurements of Edmonds (1962), which are now known to be erroneous (see Section 2.5.3), so his derived source function distribution is invalid. However, Wilson's results do serve to demonstrate the effects of geometrical smearing: near the limb, the line of sight passes through several horizontal structures whose cumulative positive and negative contributions tend to cancel out the fluctuations. The cancellation is the more complete the smaller the structure and the greater its vertical extension. This leads to a systematic reduction in the observed rms brightness fluctuation at the limb, as we noted in Section 2.5.3.

A different approach was developed by Edmonds himself in 1964. Instead of prescribing a sinusoidal distribution for the horizontal fluctuation, he chose a gaussian frequency distribution and took the ensemble average in order to derive the rms quantities. The formulation is rather opaque; he did not attempt to model realistically the dependence of the ensemble average on the heliocentric angle. Indeed, the cancellation effects at the limb, clearly pointed out by Wilson, are ignored in the approximation favoured by Edmonds.

Wilson's sinusoidal horizontal variation was used later by Altrock (1976) and by Keil & Canfield (1978) to analyse more recent centre-to-limb measurements of brightness fluctuations. They both derived a vertical distribution for the temperature perturbation very similar to those required by centre-of-disk observations in lines of various strength (see Fig. 5.2(*a*)).

These models allow the specification of the various physical quantities in the individual columns without regard to considerations of momentum or energy. As discussed in Section 4.4.4, radiative relaxation in the solar atmosphere is very rapid, much more rapid that the typical convective time scale; so we should expect that the condition of radiative equilibrium will be approximately maintained except in the deepest layers where convective energy transport contributes to the energy balance. Departures from radiative equilibrium have been sought explicitly by Wilson (1969a, b). Following the formulation originally proposed by Giovanelli (see Section 4.4.3), he used a two-dimensional Eddington approximation to solve for the mean intensity J and source function S from

$$E = -\frac{1}{3\kappa}\nabla\cdot\left(\frac{1}{\kappa}\nabla J\right) = S - J, \tag{5.2}$$

where E is the departure from radiative equilibrium: E was chosen to reproduce the centre-to-limb behaviour of the rms continuum brightness. This procedure is precisely equivalent to specifying the distribution of the temperature (source function) fluctuations and using the Eddington approximation to calculate the mean intensities and, hence, the departures from radiative equilibrium - as in an earlier analysis by Wilson (1964).

The only difference between Wilson's two treatments lay in the choice of functional forms for the horizontal and depth variations. Since Edmonds' data are now known to be erroneous, Wilson's detailed results are of little significance, but the model had the expected characteristic of an rms temperature fluctuation of 600 K just above $\tau = 1$, dropping to 200 K at a height of less than 100 km. The non-radiative (convective) contribution to the heat flux was found to drop sharply from 90% to 20% within some 20-30 km around $\tau = 1$.

A similar conclusion was reached by Turon (1973) using a different approach. Starting with a set of columns each in hydrostatic equilibrium and each having a scaled mean temperature distribution, he used the Eddington approximation to calculate the relaxation of the temperature structures with time until the system was evolving more slowly than a granule. At this point - and further evolution would remove all the perturbations since no permanent heat sources were prescribed - the temperature fluctuation was about 700 K at $\tau = 1$, falling to 150 K at 100 km and above. These estimates place upper bounds on the temperature fluctuations that can be maintained in a quasi-equilibrium state when no sources are present in the atmosphere.

None of these empirical analyses takes account of the pressure distribution and any flows that may develop. However, Margrave & Swihart (1969) have pointed out how sensitive empirical models are to the assumptions made regarding the pressure. The condition of horizontal pressure balance usually imposed is inconsistent with hydrostatic equilibrium, yet the observed vertical velocities are no larger, indeed are smaller, than the horizontal velocities (see Section 2.5.4). If hydrostatic pressure balance is maintained at all points, allowing horizontal imbalance, the densities in the various columns, and hence the opacities, are quite different. In this case the temperature fluctuations at the same geometric depth can be less than those at the same optical depth. In the horizontal pressure balance models the opposite is true. Accordingly, Margrave & Swihart concluded that a more rigorous treatment of the dynamics was required.

This contention gains additional weight if we pause for a more critical look at the assumption, underlying the previous analyses, that only temperature variations are responsible for the brightness fluctuations in Fraunhofer lines. It is well known that the profiles and, in particular, the strengths of lines formed in the solar photosphere cannot be reproduced by an atmospheric model involving only the thermodynamic quantities temperature, pressure and density. An unresolved small-scale velocity field that provides Doppler shifts varying rapidly along the line of sight must be invoked. This so-called microturbulence is usually treated as a free parameter. The subject is a real hornet's nest and possesses a vast literature (see Canfield & Beckers, 1976); we have neither the space nor occasion to provide a critical survey, but we should note that possible variations of the strength of the microturbulence between granules and the intergranular lanes could contribute to the brightness fluctuations, here attributed simply to temperature differences.

The variation between granules and intergranular lanes in the half width and equivalent width of Fraunhofer lines has been measured by several authors (e.g. Howard & Bhatnagar, 1969; Sobolev, 1975). Howard & Bhatnagar interpreted. their results as implying an increase of microturbulence of 0.4 km s^{-1} in the intergranular lanes, but only on the assumption of a 100 K temperature difference. A self-consistent determination of temperature and 'turbulent' broadening has yet to be made (see Section 5.3).

In order to remove these uncertainties we must turn to models which seek to achieve a self-consistency lacking in purely empirical analyses.

5.2.3 Semi-empirical inhomogeneous photospheric models

A fully consistent hydrodynamic description of the solar atmosphere presents many difficulties. One way of evading these is to replace mathematically inconvenient terms by parametrized forms. These parameters are then sought not by solution of the equations but by comparison with observation. Such models can be described as semi-empirical.

In this category, the model developed by Nelson & Musman (1977) deserves special consideration. They assume the granulation to be a steady flow pattern and look for solutions of the time-independent anelastic equations. Furthermore, they limit the problem to a single mode. As explained in Section 4.3.2 this can be achieved only by modelling the non-linear terms in the momentum equation. Nelson & Musman describe the coupling to turbulent scales via the inertial term as a turbulent drag, replacing the inertial and viscous terms by

$$\mathbf{F}_{\text{turb}} = -\bar{\rho}\frac{\langle v' \rangle}{L}\mathbf{v}'. \tag{5.3}$$

This can be obtained from the inertial term of Eqn. 3.24 simply on dimensional grounds. The physics of the turbulent interaction is subsumed into a single quantity L which is regarded as a scaling parameter. This parametrization is non-linear, so a bimodal planform without self-interaction can be chosen and the energy equation linearized without losing non-linear effects in the momentum equation. The radiative heat transport is treated in the Eddington approximation, assuming a grey opacity equal to the continuous opacity at 500 nm. With a final limitation to two-dimensional Cartesian geometry, the perturbation equations become

$$\frac{\mathrm{d}}{\mathrm{d}x_3}(\bar{\rho}v_3') = k_1\bar{\rho}v_1' \tag{5.4}$$

$$p' = -\bar{\rho}(v_3'^2 + v_1'^2)^{1/2}\frac{v_1'}{k_1 L} \tag{5.5}$$

$$\frac{\mathrm{d}p'}{\mathrm{d}x_3} = \rho' g - \bar{\rho}(v_3'^2 + v_1'^2)^{1/2}\frac{v_3'}{L} \tag{5.6}$$

$$v_3'\beta = \frac{4\pi}{C_p}\frac{\overline{\kappa f^2}}{\rho}(J' - S') + \frac{4\pi}{C_p}\frac{\overline{\kappa f}}{\rho}(\bar{J} - \bar{S}) \tag{5.7}$$

$$p' = (\bar{T}\rho' + \bar{\rho}T')R_* \tag{5.8}$$

$$\frac{\mathrm{d}^2 J'}{\mathrm{d}x_3^2} - \overline{\frac{f^2}{\kappa}\frac{\mathrm{d}\kappa}{\mathrm{d}x_3}}\frac{\mathrm{d}J'}{\mathrm{d}x_3} - \left(k^2 + 3\overline{f^2\kappa^2} + \overline{\frac{f}{\kappa}\frac{\mathrm{d}f}{\mathrm{d}x_3}\frac{\mathrm{d}\kappa}{\mathrm{d}x_3}}\right)J'$$
$$= \overline{\frac{f}{\kappa}\frac{\mathrm{d}\kappa}{\mathrm{d}x_3}}\frac{\mathrm{d}\bar{J}}{\mathrm{d}x_3} + 3\overline{f\kappa^2}(\bar{J} - \bar{S}) - 3\overline{f^2\kappa^2}S'. \tag{5.9}$$

Note that the horizontal averages involving opacity fluctuations must be calculated explicitly since they cannot be treated as simple linear functions of the temperature fluctuation T'.

The source function fluctuation was taken to have the LTE value

$$S' = (4\sigma/\pi)\bar{T}^3 T'. \tag{5.10}$$

The zeroth-order quantities, which are not the mean quantities since the perturbations are not linear, were evaluated from the HSRA atmosphere.

The perturbation equations were then solved numerically under appropriate boundary conditions:

(1) the velocity and mean intensity fluctuations vanish at the top of the atmosphere ($\tau = 10^{-4}$), and
(2) at the base ($\tau = 25$), the mean intensity fluctuation is equal to the source function fluctuation (the LTE diffusion approximation, see Section 4.4.3).

The magnitude of the temperature fluctuation at the base, as well as the choices of L and k_1, remain free. Nelson & Musman chose $k_1 = 2\pi/750\ \text{km}^{-1}$ after Namba & Diemel (see Section 2.3.3). The amplitude of the temperature fluctuations at the base was determined by requiring the rms brightness fluctuation at the surface to be 11% (a somewhat low value - see Section 2.5.2); the resulting temperature amplitude at $\tau = 1$ was ± 525 K (rms, 370 K). A value of $L = 210$ km (about two pressure scale heights in the atmosphere) was required to make the calculated vertical velocities consistent with the observed velocities. The *predicted* height dependences of the rms velocity and temperature fluctuations, shown in Fig. 5.3, are very similar to the empirical results in Fig. 5.2(*a*). Note that in the visible layers of the atmosphere the horizontal velocities are almost twice as large as the vertical velocities, in agreement with the available observational data described in Section 2.5.4.

The model also predicts a temperature reversal and a *downward* convective heat flux at 110 km. This is due to the very rapid radiative cooling, which is somewhat exaggerated since only continuous opacity sources are considered. The thermal inertia of the stream is very low, and the temperature excess is rapidly lost. Thereafter, any upward displacement in the subadiabatically stratified atmosphere produces a temperature lower than the surroundings. Hence the pressure fluctuation must maintain the upward flow in the face of turbulent drag and negative buoyancy, as well as drive the horizontal flow required by continuity.

At the base of the atmosphere and below, the pressure fluctuations are small and the buoyancy is balanced by the viscous drag, i.e.

$$\bar{\rho} v_3'^2/L \sim g\rho' \quad \text{or} \quad v_3'^2 \sim gL(\rho'/\bar{\rho}). \tag{5.11}$$

If L does not change significantly with height, its choice influences only the amplitude of the velocity fluctuation; the velocity scale height is determined by that of the relative density fluctuations. On the other hand, in the atmosphere the pressure gradient cannot be neglected as it serves to maintain the motion against the negative buoyancy and the viscous drag. Nevertheless, the vertical velocity distribution was found to be almost independent of L, its calculated scale height being about 160 km.

This model incorporates much of the essential physics of the granulation and may provide a more reliable estimate of the horizontal velocities in the granulation than does observation. A surprising feature of the solution is that the horizontal velocity shows no tendency to change sign – the circulation is not closed at the lowest level considered.

The model has three shortcomings:

(1) it allows no time dependence and hence throws no light on the evolution of the granulation and the processes responsible for its origin. By the same token, the production of gravity waves in the stably stratified layers is suppressed (see Section 3.4.4);
(2) it imposes a 100% correlation between brightness and velocity everywhere, whereas observation shows this not to be the case in the upper photosphere (see Section 2.5.5);
(3) it contains three free parameters which must be chosen by comparison with observations. In this sense the model is not predictive and cannot be applied to any star other than the Sun (see Section 5.5).

Fig. 5.3. Rms fluctuations in the solar atmosphere due to convective overshoot, taken from the semi-empirical model of Nelson (1978). Temperature is given in units of 10^4 K, velocities in units of km s^{-1}, and pressure in units of 10^3 N m^{-2}.

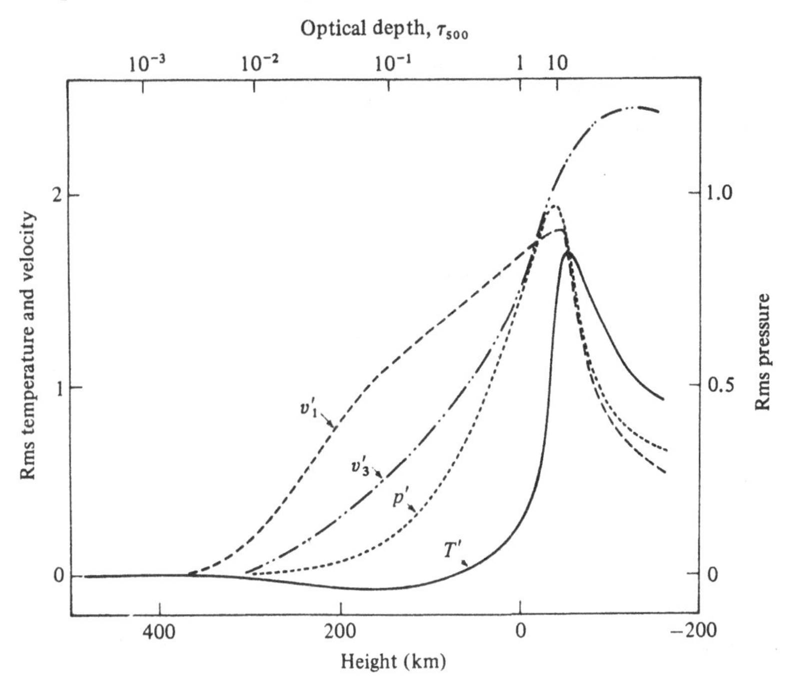

A slightly modified and improved version of this model, incorporating ionization in the energy equation and utilizing a Rosseland mean opacity rather than the continuous opacity at 500 nm, was subsequently employed by Nelson (1978) to obtain a constant-flux model solar atmosphere. The requirement that the total flux, radiative and convective, be constant renders the arbitrary specification of the temperature fluctuation at some level unnecessary; so this model has but two free parameters, L and k_1, given the total flux. The mean model was in this case adjusted for self-consistency. Nelson's model, shown in Fig. 5.1, represents the best semi-empirical description of the lower solar atmosphere to date. The mean temperature stratification differs significantly from that predicted by the zeroth-order mixing-length model, particularly in the layers where the convection is least efficient and the superadiabatic gradient highest.

5.2.4 *Theoretical granulation models*

The semi-empirical model represents a great step forward in the description of solar granulation dynamics. But the simplifications on which its tractability depends can be judged only by a more thorough analysis of the dynamics of the solar atmosphere – a much more formidable undertaking. We are fortunate today in having two numerical investigations of the dynamics of the granulation that attempt to incorporate the various hydrodynamic processes in a realistic manner.

The first is a two-dimensional simulation carried out by Cloutman (1979a) of the Theoretical Division, Los Alamos Scientific Laboratory. He investigated the *full* time-dependent equations (see Section 3.2.2) using a numerical method – the implicit continuous-fluid Eulerian (ICE) scheme. Although this method allows the presence of acoustic waves, the stability provided by the implicit formulation allows spatial and time steps large enough to follow the convective flow economically, as in the anelastic approximation.

A fine spatial mesh was chosen, with a 15 km spacing in the horizontal direction and 6.65 km in the vertical. All motions on a scale smaller than this grid spacing were modelled by a turbulent viscosity. The radiation transport was treated in the diffusion approximation (see Section 4.4.3). The diffusion of both the energy and velocity had to be artificially enhanced in order to improve numerical stability; nevertheless, the computations still effectively resolved the motions down to the grid scale.

Numerical models are necessarily of finite extent, and Cloutman chose to model the motions within a cell 173 km deep and 450 km across. The upper third of the cell included the lower photosphere; the rest was located in the upper layers of the convection zone. The total vertical extent comprised about one pressure scale height. The vertical boundaries were taken to be planes of symmetry, i.e. all gradients normal to the boundaries and the normal velocity components were made to vanish. The upper and lower boundaries were assumed to be free surfaces. The solar flux was prescribed at the upper boundary, and the temperature at the lower boundary was held constant at 10 000 K.

One could fault many of the details. The choice of two-dimensional modelling does not allow turbulent cascade to smaller scales. As we saw in Section 3.4.3, the vorticity is then conserved. The use of an eddy viscosity to describe sub-grid scales allows potential energy released by buoyant rise to be distributed as kinetic energy only to the resolved flow, i.e. the turbulence is underestimated, and the large-scale mean flows are overestimated. The boundary conditions are particularly questionable and can be justified only if it can be shown that they have no significant influence on the results, and such evidence was not provided. Whilst these objections have force, we should recognize that such modelling has been successful in describing processes susceptible to closer investigation, such as nuclear fireballs in the terrestrial atmosphere. In both this case and the granulation, the flow can be treated as essentially two-dimensional, so that at least 'semi-quantitative agreement' is expected.

It is well worth examining the physical processes in the numerical model in some detail with the aid of Fig. 5.4. The model reveals the development of a 'granule' almost filling the cell, having a diameter of 340 km. The flow is almost symmetrical, being disturbed only in the vicinity of a small subsidiary granule to one side. The large-scale flow and density distributions are sketched in Fig. 5.4; the centre line is the central axis of the main granule.

Cloutman started from a mixing-length model of the upper portion of the solar convection zone. Such models possess a slight density inversion in the uppermost convective layers whose explanation lies in the high opacities which arise before the onset of ionization. These force the temperature gradient to be very steep because the densities are too low for the convection to be efficient. The pressure cannot follow suit as the weight of the overlying gas is too small, so the density is forced to drop sharply and, in Spruit's (1974) model (see Fig. 5.1), displays a minimum some 60 km below $\tau = 1$. This tendency has been mentioned earlier to explain the choice of the pressure scale height, which is always positive, as a measure of the mixing length. A static stratification with a density inversion is Rayleigh–Taylor unstable (see Section 3.2.3), but here we have an inversion in the mean stratification of a dynamic system. It reflects only a region of very high buoyancy and occurs initially some 60 km above the bottom of Cloutman's cell. Once released from rest, the layer immediately breaks up and forms a buoyant bubble which begins to rise.

After the transients have passed, the situation shown in Fig. 5.4(*a*) results. A vortex has formed centred on the layer where the instability was originally greatest, but the low density bubble has risen, compressing material above and to the sides. 13 s later (Fig. 5.4(*b*)) the bubble has penetrated the atmosphere,

Fig. 5.4. Schematic development of a granule in the numerical simulation of Cloutman (1979a). Each panel shows one half of an approximately symmetric granule about 340 km across. The left-hand panels show the direction of the main flow and the right-hand ones the density distribution (heaviest shading indicates greatest density). Elapsed time is shown at upper centre.

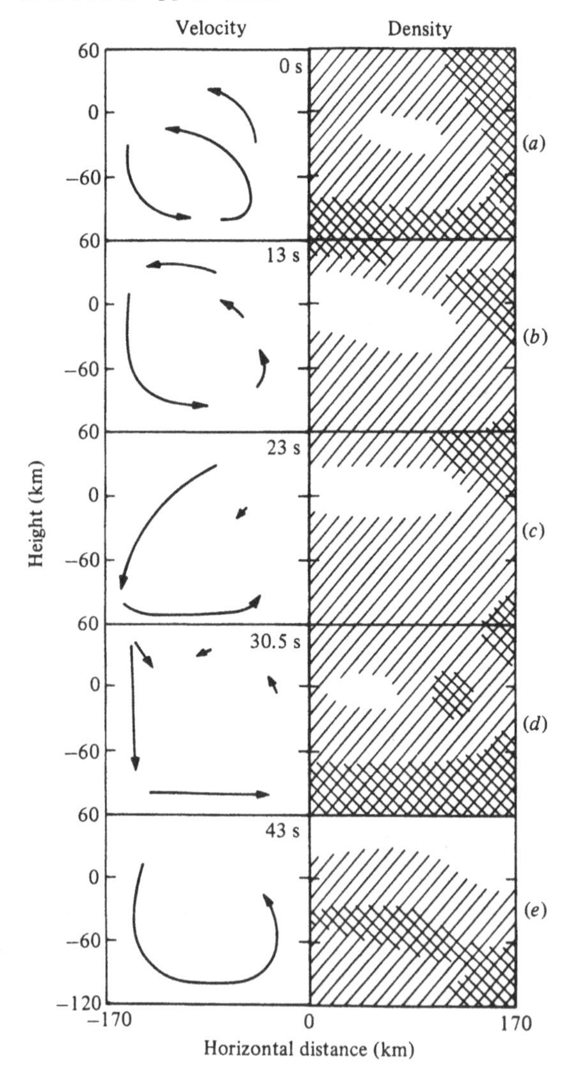

the vertical flow has stopped and a strong horizontal flow has set in which sweeps the dense layer above the bubble to the edges of the flow. After 23 s (Fig. 5.4(*c*)) the mass excess is in the process of being cleared by a strong down-flow at the boundary; there is almost no upflow anywhere. This continues for some 10 s until an almost monotonically stratified cell appears (Fig. 5.4(*d*)), with two small remnant bubbles of low and high density. Meanwhile the vortex circulation is getting under way again as the heat is dammed back by the gentle gradient, and a low density bubble again begins to form, resulting in a buoyant rise that is apparent after 43 s (Fig. 5.4(*e*)). This general process was followed through three cycles. The typical upflow velocity was 1.5 km s^{-1} and the temperature difference between the cell centre and edge was 300-400 K.

The 'stationary' state is characterized by a lower mean temperature gradient than the original; the compressible two-dimensional convection is more efficient than the mixing-length approximation would suggest. This supports the results of Nelson (Section 5.2.1). The small density inversion is eliminated in the mean state, but its incipient presence acts as a source for the bubble formation. Shears then pull the bubble into a torus (if we visualize an axisymmetric analogue) in which material rises in the centre, spreads across the top, is overtaken by the central regions of the rising torus, and is sucked back in again from below. Cloutman draws particular attention to this bubble (torus) picture, especially the detachment of the bubble from its source - the upward rise of the centre of the vortex - seen in Figs. 5.4(*c*) and 5.4(*d*). However, the gross velocity circulation does not depart strongly from that of a non-steady cellular flow. It seems arguable whether the emphasis is really warranted.

The time scale of the bubble evolution, less than 1 min, is an order of magnitude smaller than the observed granule lifetime (Section 2.3.6), whilst the horizontal scale of 340 km is a factor 2-3 too small (Sections 2.3.2, 2.3.3). In the light of the questionable assumptions, these discrepancies are not too significant, especially as the simplification to two dimensions does not allow the appearance of the granule as viewed from above to be modelled realistically; thus its evolution cannot be compared with observation. Cloutman suggests that the lifetime might be extended by regarding the first three cycles as a series of fluctuations of a single granule, only the fourth leading to new granule formation.

But granular fragmentation is most likely to be the result of a three-dimensional process, so it is doubtful whether any two-dimensional model would permit a meaningful comparison with observation.

What cannot be gainsaid is that the physical processes in the granulation deviate markedly from those embodied in the mixing-length prescription, and granulation observations should not be used to calibrate such phenomenological theories.

For a full three-dimensional simulation we must turn to Nordlund (1982). He starts from the anelastic equations, which suppress pressure waves. This approximation has already been justified (Section 3.2.3). Then a multiple horizontal-mode expansion is made retaining explicitly eight modes in each horizontal direction (no symmetry properties are assumed)

$$X = \sum_{m,l=1}^{8} (X_{cl} \cos lk_c x + X_{sl} \sin lk_c x)(X_{cm} \cos mk_c y + X_{sm} \sin mk_c y) \tag{5.12}$$

within a cell 3600 km across, i.e. $k_c = 2\pi/3.6\,\text{Mm}^{-1}$. This is equivalent to a grid of 16×16 equidistant points with a spacing of $\Delta x = \Delta y = 225$ km. The cell was chosen to be wide enough to allow the free development of all but the very largest granules (Section 2.3.3). The structure in the vertical direction and the time development were obtained by explicit numerical integration of the partial differential equations. A vertical grid covering some 1500 km using up to 32 points with a spacing of 50 km was employed. Since this is a three-dimensional model, kinetic energy cascade from large to smaller scales proceeds via the non-linear terms of the equations of motion. It is therefore necessary to remove this energy from the smallest resolved grid scales, which are still some nine orders of magnitude greater than the true viscous microscale – of the order of 1 cm in the solar atmosphere.

The expansion of the microscale to the grid scale, using a realistic approximation for the sub-grid-scale motion in the momentum and, to a lesser extent, in the energy equation, is even more critical in two-dimensional modelling. Nordlund employs a drag term similar to Nelson & Musman's (see Section 5.2.3) for the horizontal velocity components and a turbulent viscosity (see Section 4.3.1) for the vertical velocity components. The diffusion of numerical errors due to the mathematical difference scheme is not discussed. The bulk of this work remains to be published, so it is not yet possible to provide a complete analysis of the results and methods.

Nordlund's choice of periodic functions to represent the horizontal variation imposes a 'transmitting' boundary condition at the sides of the cell. Cloutman, on the other hand, considered an isolated cell with reflecting side walls. Both choices are reasonable and, probably, do not seriously influence the nature of the solution. But the conditions imposed at the upper and lower boundaries are more problematical. Nordlund treats an open system, i.e. one allowing free flows into and out of the cell. The mass exchange, especially the inflow, is controlled by the conditions outside the volume considered, and thus more or less arbitrary conditions must be imposed at the boundary. In order to minimize the influence of the material outside the box, Nordlund chose to make the vertical gradients of the velocity vanish at both boundaries. The thermodynamic boundary conditions are well formulated: the pressure fluctuations are required to vanish at the lower boundary and to follow a damped mode consistent with a superposed isothermal atmosphere at the upper boundary.

The strength of Nordlund's investigation is his formulation of the energy equation. The radiative transfer receives a realistic, though necessarily simplified, treatment which takes account of:

(1) full three-dimensionality, whereby the mean intensity is determined not by the Eddington approximation but by explicit averaging over bundles of rays;
(2) both the continuous and line absorption at no less than 368 wavelength points. Line absorption is important in the upper photosphere, since it hinders the loss of radiation and increases the cooling time (line blanketing).

He found that a spatial resolution of 50 km in the vertical direction was required for an adequate treatment of the radiative transfer.

The energy equation requires further the specification of the enthalpy of the gas entering the cell at both the upper and lower boundaries. At the lower boundary it was fixed by the requirement that the convective flux be equal to the nominal solar flux. Above the upper boundary, Nordlund assumed that the radiative cooling, or energy exchange, time is longer than the time required for the material to circulate and hence that the enthalpy is conserved. This can be enforced by making the enthalpy of the gas entering at the upper boundary equal to the average enthalpy of that moving out through the upper boundary.

Results are available only for earlier simulations. Dravins, Lindegren & Nordlund (1981) describe a calculation with 16 depth points (100 km mesh) and 3 wavelength points which did *not* include the extensive line absorption in the upper photospheric layers. The computations simulated 2 hr of granule evolution. Even after this interval, no statistically steady state was achieved, in the sense that the average values of the atmospheric fluctuations were still changing. The sequence began with the development of a strong granule which 'exploded', i.e. formed a bright ring. This removed much of the excess energy from below the surface and a 40-min quiet period followed during which only small and weak granules formed. The overall picture of development resembles that of the granulation to a certain degree and, in particular, possesses the following features:

(1) a topological asymmetry, with bright granules and dark intergranular lanes;

(2) a smaller velocity amplitude of the rising granules than that of the sinking material between them;
(3) a tendency for granules to expand until they fragment or merge.

(1) and (3) agree with observation. Comparison of (2) with observation is hindered by the difficulty of measuring reliably individual granule and inter-granule velocities, but all theoretical models of stratified convection have this property.

Fig. 5.5 shows a section through a well-developed (simulated) granule at three instants of time. Initially, a hot, buoyant bubble is seen rising and producing a cool, dense compressed region above it. Four minutes later the upward motion has ceased due to the pressure accumulation, which drives a strong horizontal flow and broadens the granule; a strong downdraft has appeared below the denser region at the edge of the granule. After six minutes the region above the granule centre ($\tau = 1$) has cooled; the temperature has dropped and the density increased at the centre of the greatly extended torus.

The radius of the toroidal vortex is greater and the vortex itself is located much deeper than in the two-dimensional calculations of Cloutman. This is not surprising as, on the one hand, Cloutman's small cell size precludes such large granule formation and, on the other hand, Nordlund's vertical spatial resolution is insufficient to pick up the effect of the narrow density inversion. The major driving force in Nordlund's model is provided by the generally convectively unstable stratification; note again that there is little or no tendency for a closed circulation to develop even at a depth of 1100 km, at which point the flow has travelled some three pressure scale heights below $\tau = 1$! No doubt the longer time scale of Nordlund's events is due to the fact that the granulation is more deep-seated than Cloutman assumed.

The rms variation of the vertical velocity component at $\tau = 1$ is about 1.5 km s^{-1} and of the horizontal component 2.5 km s^{-1}. The temperature fluctuations are not reported but the predicted rms brightness fluctuations are given as 20–30%, which are significantly higher than the observed values (Section 2.5.2).

A more detailed assessment must await the full publication of Nordlund's work, but it is already clear that, despite significant progress, neither he nor Cloutman has produced a definitive model of a granule. The dynamics of the granulation is extremely complicated, being controlled not only by the large-scale structures examined by Nordlund but also by the details of the stratification stressed by Cloutman. Further progress requires a repetition of the three-dimensional simulation with still finer grid spacing, a goal which may be within our reach with the coming generation of computers.

Fig. 5.5. Granule development in the numerical simulation of Nordlund showing the velocity vectors and isotherms in a vertical section through the centre of the granule. The isotherms are labelled in units of 1000 K. The middle and lower frames are 4 and 6 min respectively later than the upper. A few minutes later the exploding granule fragments (Dravins *et al.*, 1981).

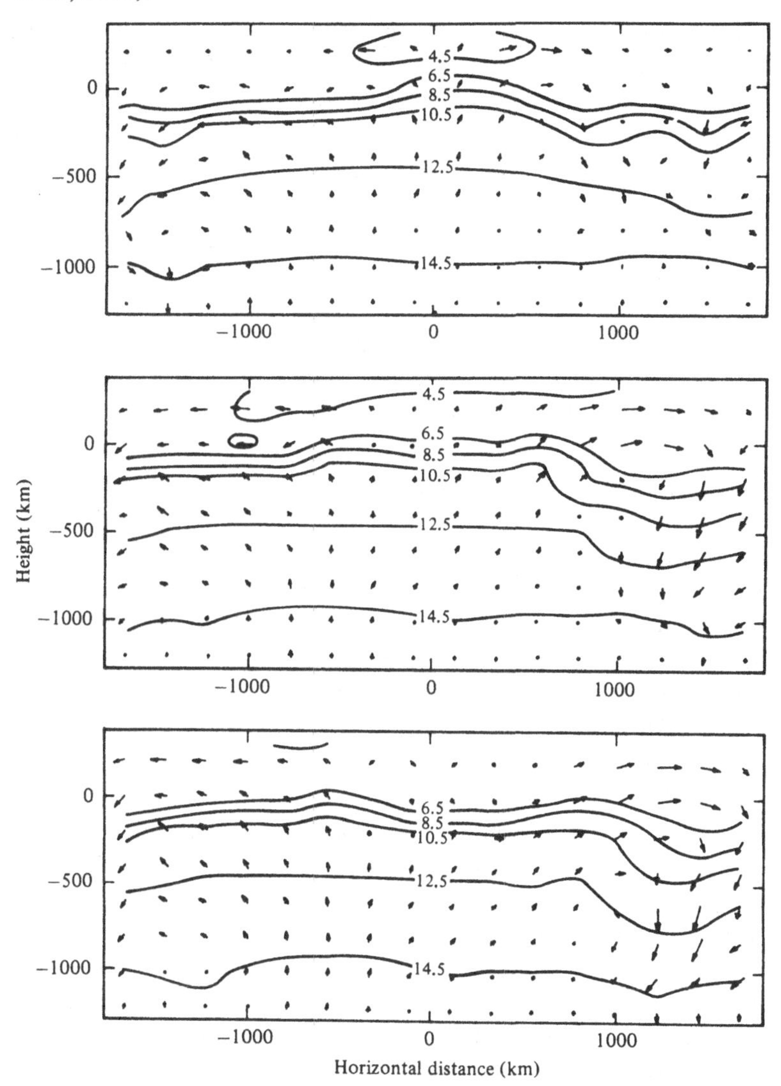

5.3 Interpretation of observed properties of the granulation

Before looking at specific properties of the granulation in the light of the theoretical insight provided by the models described in the previous sections, it is worth recapitulating some generalities.

We have already seen in Section 3.4.4 that the low atomic viscosity of the solar gas leads, in the convection zone, to vanishingly small Prandtl numbers and enormous Rayleigh numbers. The convection is therefore turbulent and, in fact, the motions become more chaotic and isotropic *the smaller their spatial scale*. These turbulent motions are parasitic: they drain kinetic energy from the large-scale motions driven by the convective instability and feed it into microscales that even the weak viscosity can dissipate. Only at scales far removed from those of the driving motions is it sensible to seek fully developed turbulence and such scales are unobservable as resolved motions; they may at most contribute to the 'microturbulent' regime of spatially unresolved motions.

The notions of isotropy, homogeneity and randomness do not apply to the motions which *produce* the turbulence. Here we can expect to find a large-scale structuring, a quasi two-dimensional flow pattern. There is no contradiction in the picture of large-scale ordered motion coexisting with small-scale turbulence. The buoyancy forces favour the establishment of large-scale flows but they are balanced by inertial forces which convert the flow into ever smaller scales. Hence the characteristic feature of turbulent cellular convection is its non-steady nature. A cellular flow is no sooner generated than it is disrupted, only to reappear with similar qualitative but differing quantitative properties. Order appears but then relapses into disorder. The interplay is fundamental to non-linear theories of solar and stellar convection in general and of the solar granulation in particular.

What we see as the granulation is the instantaneously ordered large-scale pattern. The organization is imposed by the need to carry heat, hence the granulation is a direct manifestation of the convective heat transport. The turbulent scales carry almost no convective energy and contain far less kinetic energy. Nevertheless, the dynamics of the larger scales cannot be understood without reference to the smaller scales, and one task of theory is to describe the conversion of the ordered motion into turbulence. The correctness of the description can be judged only by comparison of its predictions with observations of the *ordered* scales. In the following we shall examine to what extent the properties of the cellular-like flow that we call the solar granulation can be understood within the framework of present-day theories and models.

(1) *Mean cell size*. The granulation pattern has a well-defined and rather narrow distribution of cell sizes (Fig. 2.10), the mean value being 1400 km (Section

2.3.3). Linearized theory gives a growth rate curve that shows no maximum that can be identified with the peak of the observed cell size distribution; an explanation lies within the realm of non-linear theory.

Semi-empirical models require the *adoption* of a mean cell size and hence make no prediction. Nevertheless, the mechanisms controlling the order of magnitude of the size of the surface convection are indicated by such models. Nelson & Musman (1978) used their steady-state two-dimensional model to investigate the behaviour of cells with different horizontal sizes.

If the horizontal size is less than 100 km there is insufficient optical thickness between the hot rising and cool falling currents to prevent rapid radiative energy exchange, which then damps out the temperature fluctuations below the visible surface. Since the buoyancy is lost, such motions do not reach the surface. If the cell is too large, it acquires insufficient energy in passing through the upper convective layer to drive the horizontal flow. The pressure build-up is then sufficient to bring the motion to a halt. This implies a limit of some 3300 km on the size of the superficial convection cells. Note that this does not rule out deep-seated convection such as we suppose the supergranulation to be (see Section 5.4).

An immediate corollary is that large granules require a large pressure gradient between the centre and boundary in order to drive the horizontal flow. The larger pressure at the centre increases the continuous opacity and raises the mean level from which the radiation escapes. Since the temperature decreases with height, with sufficient pressure the temperature drop associated with this increase in the effective height of the emission partially compensates the local temperature difference between the hot and cold regions. The centre can then look dark. Nelson & Musman remark that the dark spot observed in the centres of some granules (see Section 2.3.7) may in fact cover the hottest part of the granule. They also claim that the dark spot may be associated with an upflow and not a downflow - this would reduce the brightness-velocity correlation - but time-dependent modelling shows this to be an oversimplification.

In principle, numerical simulations should demonstrate the preferred scale if the computation is carried out over a large enough volume for the artificial boundary conditions to have little influence. This condition seems to be only marginally met by the simulations of Nordlund and even less so by those of Cloutman (see Section 5.2.4). The limited length of time over which the development was followed also reduces the significance of the preferred scales shown by their solutions. Cloutman's granule did not evolve beyond a width of 340 km - unrealistically small - whilst one of Nordlund's achieved a diameter of some 3000 km, at which stage it showed characteristics similar to those of an 'exploding' granule.

It is of some interest to calculate the ratio $Q = d_c/d_3$ ($= 2\pi/\tilde{k}_\perp$), where d_c is the mean cell size and d_3 is the effective thickness of the convecting layer. We may take the upper boundary of the layer to be located at $\tau = 1$ since at this level the convecting elements begin to radiate away a substantial fraction of their thermal energy. The lower boundary is not so well determined; examination of Fig. 4.1 shows that regions of large superadiabatic gradient ($\nabla - \nabla_{ad} > 0.1$) extend over some 300 km, whilst the most highly superadiabatic layer lies only ~100 km below the surface. Thus we might expect, with $d_c = 1400$ km, that

$$5 < Q < 14.$$

The models provide little further information: Cloutman's boundary conditions enforce a circulation within his box and he finds

$$Q \simeq 3.4,$$

but this value may be influenced by the restricted horizontal dimension in his simulation. The other authors allow flow across the lower boundary. A closed circulation is not enforced, nor is one found. Thus we can give only upper limits to Q. The results of both Nelson (1978) and Nordlund imply an upper limit of about 3.

According to Simon & Leighton (1964), these values are close to those for stationary Bénard cells ($Q = 2$-3), whilst a value of Q ranging from 5 to 10 is characteristic of non-stationary convection observed in nature over a very wide range of densities and cell sizes. As an example, they cite Tiros photographs showing convection patterns in cloud formations for which $Q = 10$.

The discrepancy would appear to be related to the failure of the model flows to break up into small cells stacked vertically. On present evidence, it seems likely that the granulation extends significantly deeper than a pressure scale height but not so deep as to contradict our supposition that the granulation is a phenomenon restricted to the upper layers of the solar convection zone. As yet, though, little light has been shed on just how the preferred cell size shown by the granulation arises.

(2) *Lifetime*. One might expect the observed lifetime to be comparable to the time required for a fluid element to make a circuit of the convection cell. From Nelson's model shown in Fig. 5.3, we can see that the average rms convective speed is about 1.5 km s^{-1}, so the maximum speed lies somewhere between 2 and 3 km s^{-1} depending on the geometry of the cell. A simple calculation for a circuit $\frac{1}{2} \times 1400$ km across and some 450 km deep ($Q \simeq 3$) confirms this expectation. For an average granule this would require a time of the order of 800–1200 s = 13–20 min, which agrees roughly with the estimates of the lifetimes of granules that survive to fade away (Section 2.3.6).

The possibility that the granules - notwithstanding their cellular appearance - actually represent convective plumes rather than cells can be excluded. This type of convection is well known in meteorology (see, for example, Priestley, 1959 and Turner, 1973) and was mentioned by Simon & Leighton (1964) in connection with the supergranulation. In this case the travel distance is of the order of 450 km, and the time required becomes ~3 min, which is much too short for a typical granule.

Indeed, closer scrutiny of the simulated flows endorses the quasi-cellular flow picture. The underlying vortical motion defines a cell whose visual appearance is modulated by the formation and transport of low density bubbles. These rise to the surface, compressing the material above them which is then forced to flow to the sides. By the time the resulting horizontal flow is well developed the upflow has stopped, differing thereby from the stationary model of Nelson & Musman (1977). The 'granule' then lasts only as long as it takes this material to drain away, allowing another bubble to force its way up. By the time the granule can be seen to show a resurgence, the circulation is almost completed. However, observations suggest that granules which fragment have shorter lifetimes than those that simply fade away, implying that fragmentation leads to a premature termination of granule development. Two-dimensional models, which prohibit three-dimensional instabilities, tend to overestimate the mean granular lifetime. The reasons for the growth, expansion and decay of individual granules are clear, but the exact manner in which decay occurs is not. Merging would be the obvious process, a more dynamic neighbour entraining a weaker predecessor, but observation shows that fragmentation is more frequent (Section 2.3.7). The cause of the fragmentation has not yet been identified in the numerical simulations although the instability described by Jones & Moore (1979) is a plausible mechanism (see Section 3.4.3).

(3) *Granulation near the extreme limb; height of overshoot*. The granulation has been observed up to a distance of 5″ of arc from the limb. When the transfer of radiation is taken into account, all analyses of this visibility agree that the temperature fluctuations do not extend above a height of 50 km in the atmosphere. The temperature fluctuations are extremely rapidly damped by the short cooling time of the lower atmosphere. Indeed the cooling is so rapid that a single parcel of gas moving upward with a typical convective velocity would lose all its temperature excess in the visible layers. That we see as much excess as we do near the limb is due to the continued heating of the parcel by radiation arising from the hotter material still coming up from below (Nordlund, 1976; Musman & Nelson, 1976). Convective heat transfer, which requires the *transport* of a heat excess, does not extend far into the convectively stable atmosphere.

The various estimates of the height of cessation of the convective flux in the atmosphere are summarized in Fig. 5.6, which is based on a diagram of Edmonds (1974). The two shaded areas denote the predictions of mixing-length theories, the one with the steeper cutoff being the local theory and the one with the gentler cutoff the non-local theories of Spiegel and Ulrich (see Section 4.2.4). The semi-empirical two-dimensional models of Nelson & Musman (1977) and Nelson (1978) show a gradient intermediate between the two forms of mixing-

Fig. 5.6. The convective heat flux in the solar photosphere, expressed as a fraction of the total flux. The two shaded regions indicate the fluxes predicted by the (*A*) local and (*B*) non-local mixing-length theories. Observational values are identified at upper left, semi-empirical model predictions at lower right (after Edmonds, 1974).

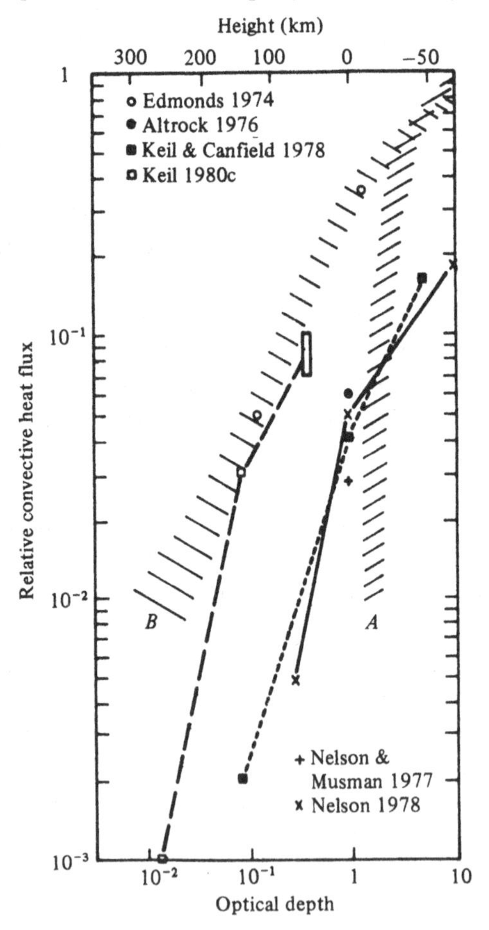

length formulation, as explained in our discussion of Section 4.2.4. The observations reflect the manifold uncertainties that have been described exhaustively in Chapter 2 and Section 5.2.2; little more need be said here.

Exceptionally large convective flux values are reported by Keil (1980c), a consequence mainly of the large convective velocities derived in his analysis (see Section 2.4.2). The results of Edmonds (1974) are anomalous, being based on smaller observed rms values, and should be treated with caution. Generally speaking, the observations favour the rate of decrease of convective flux in the atmosphere predicted by the semi-empirical models which treat the transfer of radiation realistically.

Unlike the temperature fluctuations, the convective motions overshoot well into the atmosphere (see Section 3.4.5). Although the upward-moving fluid loses its buoyancy along with its temperature excess, the momentum gathered during the acceleration phase while it is crossing the strongly superadiabatic layer is not immediately lost. However, it is progressively reduced first by the pressure gradient, then by the inertial drag due to the production of turbulence.

Nelson & Musman claim a scale height of 160 km for the vertical velocity which almost vanishes by a height of 300 km. As regards the observations of velocities, a crucial consideration is the separation of the remnant granular motion from the smaller scale motions generated by it. This separation has not yet been satisfactorily accomplished, so comparison with theory is inconclusive.

The change of sign in the temperature fluctuation above 100 km predicted by Nelson & Musman (1977) finds little support in the observations of average granule behaviour, though isolated cases were found by Altrock & Musman (1976). The time-dependent results shown in Figs. 5.4 and 5.5 indicate that it might well be a *transitory* effect.

All models predict that the larger granules, which lose less buoyancy by horizontal radiative energy exchange below the surface, should have greater speeds at the surface and hence should overshoot further into the atmosphere. At scales larger than the granulation the pressure forces needed to produce the horizontal flows increase to the point where they significantly counteract the upward motions in the subsurface layers, so that the overshoot cannot increase without limit. A sharper drop in granular vertical velocity power with height in smaller scale structures than in larger scale structures seems to be apparent in the observations (Durrant & Nesis, 1982; see the power spectra in Keil, 1980b).

The brightness structure of the upper photosphere seen in the cores of medium-strong lines (see Sections 2.4.2 and 2.5.5) has yet to be investigated theoretically. Its origin is a matter of conjecture, though several authors (e.g. Durrant & Nesis, 1981) have noted the possibility of gravity waves being induced by perturbation of the equipotential surfaces of the atmosphere as a result of convective overshoot. The properties of these waves in the solar atmosphere have since been extensively discussed by Mihalas & Toomre (1981, 1982).

(4) *Direction of cellular motion.* In gaseous convection in the laboratory the direction of circulation in a cell appears to be such that the fluid moves downward at the centre of the cell and upward at the cell boundary. Analysis indidates that this is caused by the temperature dependence of the viscosity, which increases with temperature in gases (see Section 3.4.2). In the Sun the circulation has the opposite sense. The strong stratification in the Sun appears to be the main reason for the difference. The anelastic two-mode models of Massaguer & Zahn (1980) for an A-type star convection zone show that the laboratory sense circulation is less efficient than the solar sense. The role of the pressure now needs to be considered; the pressure excess necessary to drive the horizontal motion acts in a non-linear manner to hinder and spread the upward flow whilst accelerating and funnelling the downflow. This in turn leads to more efficient heat transport. The wide column of rising material hinders the loss of excess temperature by horizontal radiative transfer. The hot material thus radiates away most of the excess at the surface, sinks and is effectively reheated by radiative transfer and turbulent mixing in the narrow dark lanes.

(5) *The turbulence spectrum.* Both laboratory experience and numerical modelling require the large-scale quasi-cellular motions to break up and be converted into turbulence. This turbulent regime has scales too small to appear as resolved motions, but we have two indirect means of estimating the total turbulent field. The first is by means of an extrapolation to smaller scales of the observed velocity spectrum. The use of the Kolmogoroff inertial law for this purpose is dubious as the law is not followed, for example, by turbulence in the terrestrial atmosphere. However, the exact wavenumber dependence assumed is not critical - the sharp drop in the kinetic energy spectrum at wavenumbers beyond the buoyancy-driven scales ensures that the total kinetic energy in the turbulent field is small. Nelson & Musman (1977) estimate the rms turbulent velocity to be about 0.4 of the resolved rms velocity, so the microturbulent component should drop rapidly in the atmosphere from a maximum value of 0.6 km s^{-1}. The second approach is through measurements of the microturbulence alluded to in Section 5.2.2. A conservative estimate of this quantity in the photosphere is 0.8 km s^{-1} from disk-centre observations and 1.2 km s^{-1} from observations in light integrated over the solar disk (Blackwell, Ibbetson, Petford & Willis, 1976), much greater than any feasible estimate from turbulent dissolution of granules.

There are two avenues of escape from this dilemma. Either the extra microturbulence is due to another form of motion - acoustic waves perhaps - or we have underestimated the vertical gradients in the large-scale granular flow. The second point of view has been strongly advocated by Nordlund (1978). With a highly simplified model of the large-scale steady flow, he correctly reproduced the widths of lines but found that the increase of equivalent width above the value in the static atmosphere was an order of magnitude too small. A more detailed steady-flow model showed a similar discrepancy, but preliminary results of a time-dependent simulation (Nordlund, 1980) indicate that allowing granules to form and dissolve increases the mean vertical velocity gradient sufficiently to explain the residual line strengthening at the centre of the disk.

To conclude this section, we emphasize once again that the interpretation of many of the observed features of the granulation lies within the scope of non-linear theory, whose detailed application to the granulation has begun but whose application to the solar convection zone as a whole is still in its infancy. In this section we have tried to confront existing theoretical ideas with just a few of the basic observational facts. The interpretation of other observed features such as granule shape, evolution, diversity in brightness, etc., must await the appearance of more realistic numerical simulations.

5.4 Interpretation of the supergranulation

There are no models of the dynamics of supergranulation cells. As already noted in Section 2.7 the evidence for a convectively driven flow is poor. The supergranulation is observable mainly as a *velocity* pattern, and only very weakly as an intensity pattern. This implies that the corresponding thermal energy convected to the surface is much smaller than in the case of the photospheric granulation. This does not necessarily mean that the supergranulation is not a convective phenomenon: the lack of evidence could simply be a consequence of the driving region being located deep in the convection zone. This is the view taken by Simon & Weiss (1968), for example.

These authors discuss in a highly simplified manner convection in a stratified layer. They point out that, in a system with a constant scale height, motions extending over several scale heights are more efficient than mixing-length eddies since they can convey the heat at a smaller superadiabatic gradient. (We have noted this behaviour in Section 4.3.) But if the scale height decreases with height, the benefit gained by extending the buoyant motion in the vertical direction will at some point no longer outweigh the amount of energy lost in driving the widespread horizontal motion. So the deep, broad cell should break up into smaller cells when the imbalance becomes critical. The authors then argue, using the example of a polytrope, that it is plausible to assume that the motion originating at a point where the scale height is H will form a cell of diameter $2H$, extending from $-0.1H$ to $-H$. On this basis, Simon & Weiss tentatively distinguished three scales of motion in the solar convection zone:

(1) a large-scale motion extending from the base of the convection zone (150 000 km) throughout the region where the polytropic index is sensibly constant, i.e. where the scale height is a linear function of depth;
(2) an intermediate-scale motion filling the region where the stratification is almost adiabatic but the polytropic index increases rapidly due to the onset of ionization, first of helium and then of hydrogen (see Fig. 4.1), between 15 000 and 1600 km below the the photosphere. The ionization of neutral and singly ionized helium as the mechanism responsible for the supergranulation was suggested earlier by Simon & Leighton (1964);
(3) a small-scale motion within the layer of high superadiabatic gradient, that is, above a depth of 1600 km.

The third member of the hierarchy is, of course, the granulation. The second they identified with the supergranulation. In the visible layers of the solar surface this mode is replaced by the granulation as the major means of transporting convective heat, so there is no further temperature fluctuation associated with the overshoot of the supergranulation velocities.

The possibility of the existence of a hierarchy of cells stacked one above the other, extending down to the bottom of the convection zone, was raised by Bray & Loughhead (1967). They speculated that the lowest regions manifested themselves at the photospheric level in motions having a horizontal scale even larger than that of the supergranulation but which, at the time, had not been observed or identified. Since then considerable effort has been devoted to the search for such 'giant cells'. Extrapolating the progression from the granulation to supergranulation, one would expect the photospheric appearance of giant cells to be weaker still as a velocity field and non-existent as an intensity field. Moreover, we should expect much longer time scales for the large-scale flow, perhaps comparable to the period of rotation of the Sun (~26 days). Such an approximate coincidence would lead to a strong coupling between the rotation and circulation and a marked distortion of the flow pattern (see various papers in Bumba & Kleczek, 1976), producing perhaps banana-shaped cells rather than round cells.

However, careful measurements of the velocity fields across the disk have failed to reveal any systematic large-scale pattern with magnitude greater than 10 m s^{-1}. A less direct indicator of organized flows in the subsurface layers is the distribution of magnetic fields. We suspect that the motions of these fields reflect those of the deeper layers of the convection zone in which they are rooted by virtue of the high electrical conductivity of the gas. The field distribution does indeed outline the supergranulation pattern (Section 2.7); furthermore, Dodson & Hedeman (1968) claim to have detected patterns of active regions centred on 'active' longitudes that may imply a subsurface flow pattern on a scale comparable to the solar radius.

An alternative picture of the supergranulation that does not invoke subsurface convection has been put forward by Cloutman (1979b), who visualizes the horizontal supergranular flow as a rip current. He argues that the granular 'bubbles' push gas up into the photosphere like a piston as well as lifting up more material from below by entrainment. The balancing downflow is inhibited by the close packing of the bubbles which leads to a steady accumulation of material in the upper photosphere. He observed such an effect in his numerical simulation of the granulation (see Section 5.2.4). Since this was a closed system, the accumulation weakened the bubble formation until the excess could drain back, whereupon a strong granule formed anew. In the open system of the solar atmosphere, the excess can be transported horizontally by pressure gradients over large distances as a rip current which, according to Cloutman, we see as the horizontal supergranular flow.

These arguments do not suggest how the scale of the supergranulation is determined. Cloutman notes that the rip current may be disrupted by the Kelvin-Helmholtz instability and estimates the wavelength of the unstable mode to be 0.2 to 1.0 times a supergranule radius. Alternatively, he suggests that the downflow may take place in the magnetic elements of the network bordering the cells. In such a case, the supergranulation scale is determined by that of the magnetic distribution and we are then faced with providing an explanation for this!

A related mechanism had been suggested earlier by Naze Tjötta & Tjötta (1974). They appealed to non-linearities in a localized vertical oscillation of finite amplitude to provide a vertical mass flux. A horizontal wave, either standing or travelling, was then necessary to transport away the accumulated material. The authors showed that the 5-min oscillations have sufficient amplitude to allow this mechanism to operate, but we are still faced with a dilemma similar to that confronting the rip-current hypothesis - what causes the global oscillations to have the peculiar structure required to produce the supergranulation?

None of these suggestions is entirely convincing. It is quite plausible that the supergranular flow acts to relieve mass accumulation in the upper solar atmosphere. But to conceive the flow as due primarily to a deep-seated convective process makes less demands on the imagination.

5.5 Granulation effects in mean solar line profiles

5.5.1 Introduction

In Chapter 2 we dealt, *inter alia*, with measurements of the spectral properties of *spatially resolved* granules. In Section 5.2.2 we met the use of the asymmetry of *spatially unresolved* line profiles in the derivation of early inhomogeneous models. The origin of the asymmetry is easy to understand. Let us assume, for simplicity, a 'two-stream' model of the region of line formation, i.e. one with alternate columns of hot upward-moving and cold downward-moving material. If the spatial resolution is such that the profiles corresponding to the hot and cold elements cannot be separated, the profile recorded on a spectrogram is a superposition of the two profiles, each shifted by an amount corresponding to its Doppler velocity. These profiles are depicted schematically in Fig. 2.17. If the 'hot' and 'cold' profiles were identical, no asymmetry would result, only a broadening. In general, however, this will not be so: not only will the continuum level for the two profiles be different but for lines of high excitation and ionization potentials, for example, the 'hot' profile will certainly be stronger than the 'cold' (Voigt, 1956). In such a case it is clear that their superposition will produce an asymmetrical profile whose violet side will be more extensive than its red.

If an accurate mean line profile is available, the locus of the midpoints of the lines joining points of equal intensity may be plotted. The resulting curve is known as the line bisector, also called from its characteristic form the C-shape of the line (see Fig. 2.17); it conveniently displays the degree of asymmetry as a function of wavelength.

A second way in which the inhomogeneity of the photosphere can influence a Fraunhofer line is by causing a displacement, the mean position of the line being shifted towards the stronger component. The measured displacement is dependent on the procedure employed, since the degree of asymmetry varies across the line. The line core does not yield the same displacement as the centre of gravity of the profile as a whole.

It is clear that these quantities are net effects due to incomplete cancellation of the positive and negative fluctuations in the residual line intensities. The mean line displacement is, therefore, much smaller in magnitude than the resolved displacements and requires great precision for its measurement. Furthermore, the information content of the mean solar line asymmetries and displacements is limited, as will be discussed below. However, the Sun is the only star on which we are ever likely to see resolved granules, and so the effects of convection on unresolved line profiles are a vital link between solar and stellar granulation.

5.5.2 *Shifts and asymmetries of mean solar line profiles*

The discovery that the wavelengths of Fraunhofer lines, corrected for solar rotation, are not constant across the disk goes back to 1907, but reliable measurements made with reference to a standard laboratory or telluric source stem only from the last 30 years. The most extensive compilation is due to Pierce & Breckinridge (1973), who list the wavelengths of 14 624 lines between 292 and 900 nm measured with respect to a hollow-cathode thorium lamp. These Kitt Peak wavelengths were obtained photographically by determining the position of the bottom 5-10% of each line, and are thus strongly weighted to the line cores.

Dravins *et al.* (1981) selected 311 lines of neutral iron which are apparently unblended and for which accurate laboratory wavelengths are available. They corrected the measured wavelengths at the centre of the disk for the relativistic gravitational redshift given by the formula

$$v_{rs}/c = GM/Rc^2, \quad (5.13)$$

where M and R are the stellar mass and radius and G is the gravitational constant. For the Sun, this yields $v_{rs} = 636\,\mathrm{m\,s^{-1}}$. The results are shown in Fig. 5.7 for lines of different excitation potential and strength. Strong lines of all excitation levels show an almost constant blueshift amounting to $200\,\mathrm{m\,s^{-1}}$. Weaker lines show larger shifts that increase with excitation potential to as much as 700-$800\,\mathrm{m\,s^{-1}}$.

As one moves towards the limb these shifts decrease. Fig. 5.8, taken from Brandt & Schröter (1982), shows the centre-to-limb behaviour of the Fe I λ557.6 nm line measured at various intensity levels - line centre, 40%, 60% and 80%. All show a slight increase in the blueshift as one moves away from the centre of the disk, and then a steady decrease at heliocentric angles greater than 45°. Close to the limb the reliability of the measurements becomes low, but the shifts appear to converge to zero for all lines regardless of strength. This is called the 'limb' effect but in reality is a centre-of-disk effect.

When the Sun is observed in integrated light (i.e. as a point source), the limb effect somewhat dilutes the line shift at disk centre and the resulting shift of the integrated profile is some $100\,\mathrm{m\,s^{-1}}$ less than the shift of the mean profile at the centre of the disk. The differential shifts in integrated light for weak lines are thus $\sim 600\,\mathrm{m\,s^{-1}}$, which is equivalent to 0.001 nm at 500 nm.

The differing shifts of lines of different strength is simply a reflection of their differing sensitivity to the presence of the granulation. The temperature fluctuations are confined to the lowest atmospheric layers and thus affect the cores of weak lines more than those of strong lines. This same height variation also gives rise to the asymmetry within a strong line since radiation at each wavelength stems from a different range of heights. Just as the blueshift increases for lines of decreasing strength, so the asymmetry increases for points of decreasing absorption, i.e. higher in the line profile. Further out in the wings of the line the asymmetry diminishes again due to opacity effects discussed below.

The resulting C-shape varies most markedly with the strength of the line, other line parameters such as wavelength and excitation potential having little effect. Fig. 5.9 shows the mean C-shapes of the Fe I lines investigated by Dravins *et al.* (*a*) measured at the centre of the disk and (*b*) integrated over the disk. The magnitude of the variation is reduced from (*a*) to (*b*), from 300 m s^{-1} to less than 200 m s^{-1}, but the characteristic form survives. The reason is clear from an examination of the centre-to-limb behaviour. Brandt & Schröter (1982) found that the line bisector of the Fe I λ557.6 nm line retains the C-shape out to a heliocentric angle of 45°, and then straightens out until, at the limb, no C-shape was evident within the accuracy of the measurements (Fig. 5.10). The contribu-

Fig. 5.7. Shifts of Fe I lines at the centre of the solar disk as a function of excitation potential of the lower level. The line depth is indicated by the size of the symbol. Shifts are expressed as apparent radial velocities relative to a laboratory standard and are corrected for the gravitational redshift (Dravins *et al.*, 1981).

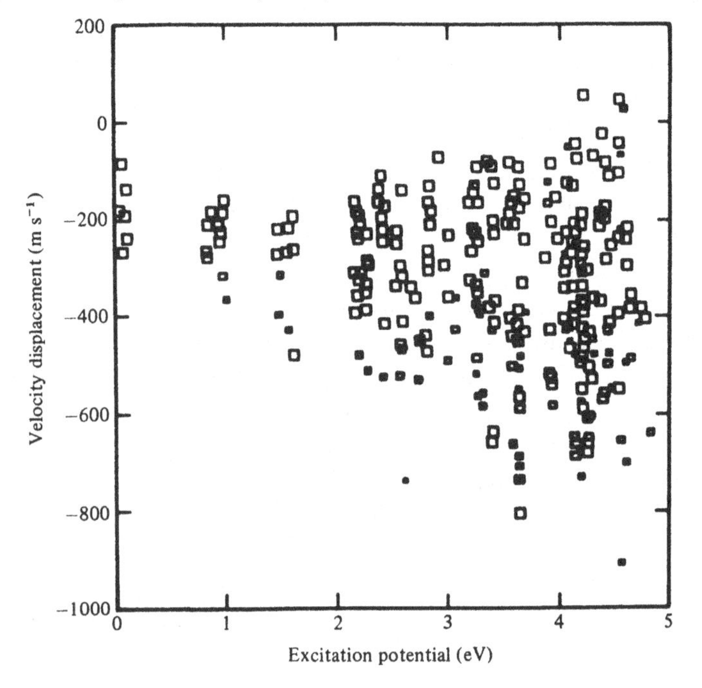

tion of the limb regions to the integrated profiles serves only to dilute the C-shape at disk centre.

The asymmetry and shift of the integrated profiles are most probably statistically stable quantities. The shift of spatially averaged but non-integrated profiles, on the other hand, varies markedly with time and position on the solar disk. Brandt & Schröter find that the average shift of the line at points along the polar diameter is quite different from those along the equator. Such global latitudinal variations had previously been suspected by Beckers & Taylor (1980). A connection with the degree of activity was reported by Howard (1971) and Ambrož (1976).

The C-shape, on the contrary, shows neither latitudinal nor temporal variation in *non-active* regions of the Sun. Especially noteworthy is the finding of

Fig. 5.8. Centre-to-limb variation of the shift of the Fe I λ557.6 nm line measured at line centre and various points of the profile, relative to the shift at disk centre. Points in the profile are defined by the value of the residual intensity, expressed as a percentage of the local continuum (Brandt & Schröter, 1982).

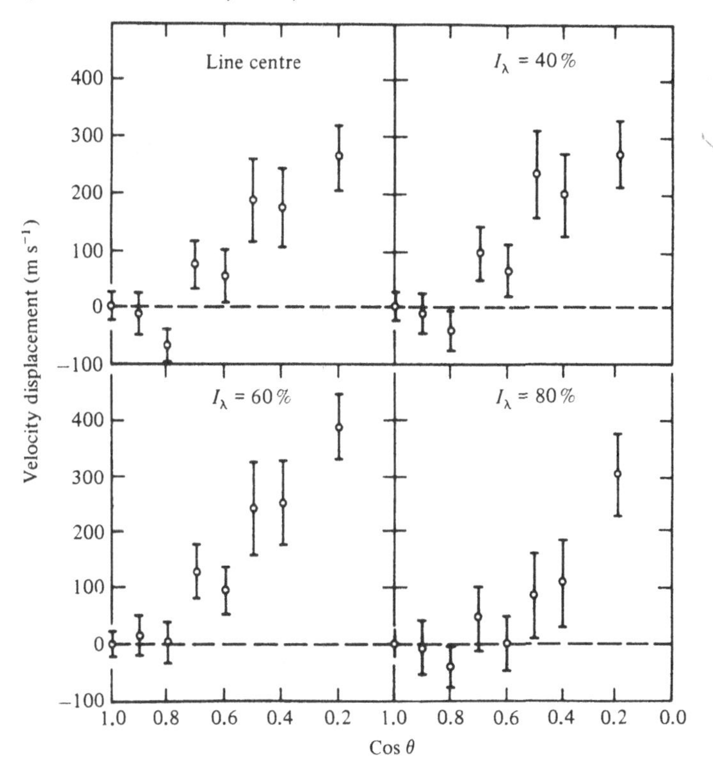

Koch, Küveler & Schröter (1979) that the C-shape is independent of the phase of the 5-min oscillation. However, the constancy of the C-shape does not extend to *active* regions, where Kaisig (1981) observed a more pronounced blueshift in the lower part of the line profile than in quiet regions. These observations

Fig. 5.9. Dependence on line strength of bisectors (C-shapes) of (*a*) 311 Fe I lines observed at disk centre; (*b*) the same lines integrated over the disk; and (*c*) synthetic lines of different strength from the granulation simulation of Nordlund (Dravins *et al.*, 1981).

The lines are grouped according to residual intensity at line centre expressed as a percentage of the local continuum.

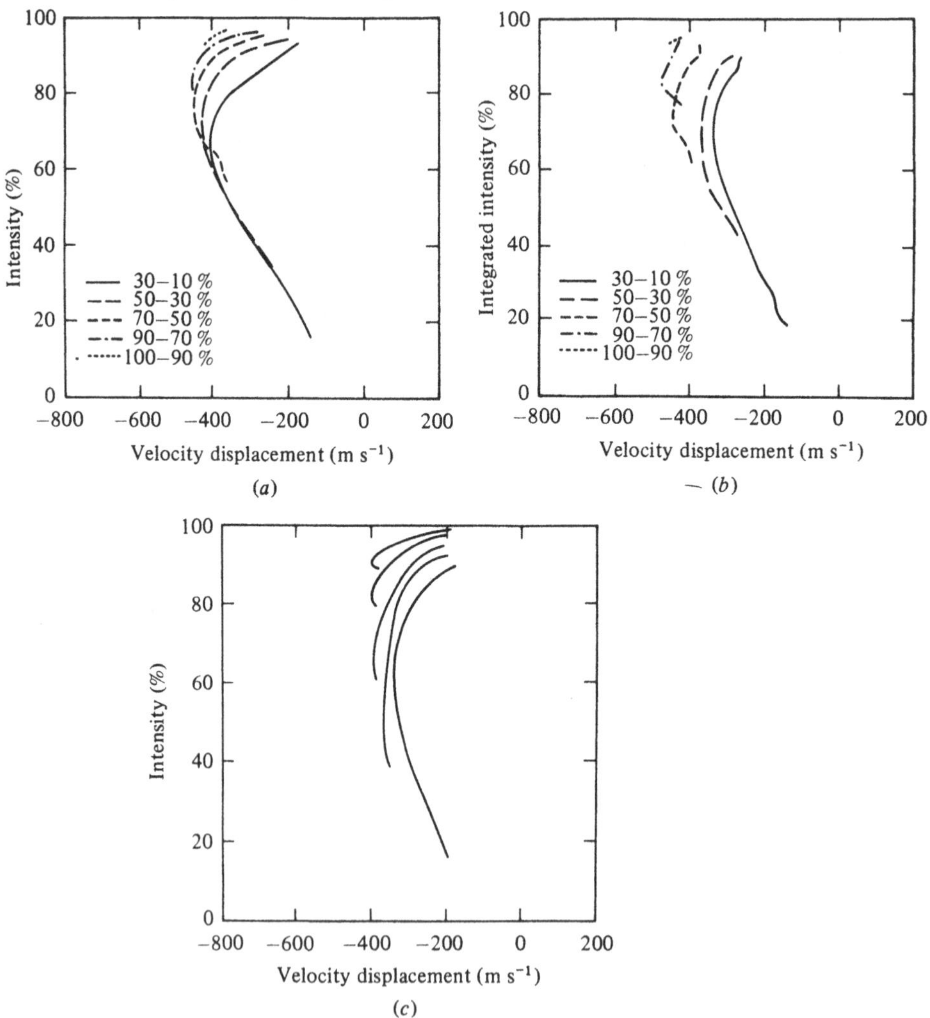

raise the possibility of a dependence of the properties of the granulation on solar activity and, perhaps, latitude.

5.5.3 *Convective origin of mean line profile asymmetry*

It is worth analysing the origin of the asymmetry in a little more detail. As in Section 4.4.1, let us look at the change in the profile due to a perturbation of the atmosphere. If we expand the expression for the emergent intensity to second order we obtain for the disk centre

$$I = \int S\,e^{-\tau}\kappa\,dx_3 + \int (S\kappa)'_1\,e^{-\tau}\,dx_3 - \int S\kappa\,e^{-\tau}(\tau)'_1\,dx_3 + \int (S\kappa)'_2\,e^{-\tau}\,dx_3$$

$$-\int (S\kappa)'_1(\tau)'_1\,e^{-\tau}\,dx_3 - \int S\kappa\,e^{-\tau}\{(\tau)'_2 - [(\tau)'_1]^2/2\}\,dx_3, \qquad (5.14)$$

where $(\ )'_1$ and $(\ )'_2$ are perturbations of first and second order in the fluctuations of thermodynamic quantities and

$$\tau' = \int \kappa'\,dx_3, \qquad (5.15)$$

etc.

Fig. 5.10. Observed variation of the bisector of the Fe I λ557.6 nm line as a function of heliocentric angle, θ. The bisectors are labelled with the value of $\cos\theta$. The displacements are expressed as apparent radial velocities; the zero point is arbitrary. For clarity error bars have been suppressed (Brandt & Schröter, 1982).

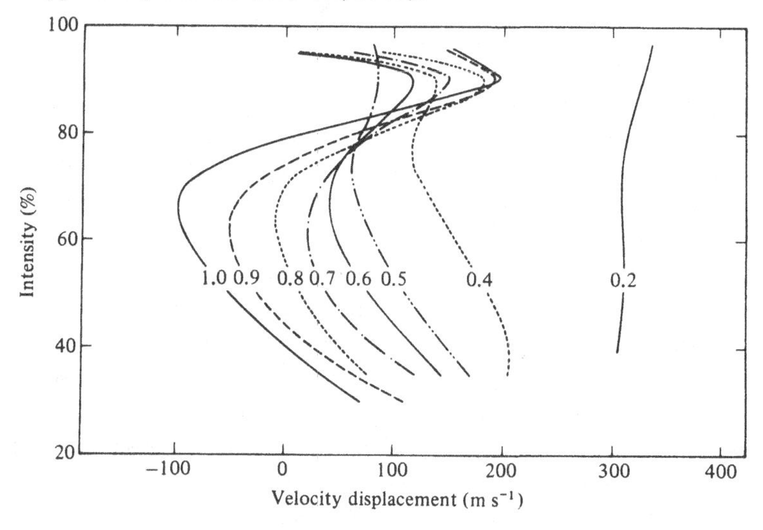

To obtain the mean profile we take the spatial average; by definition the first-order fluctuations vanish, leaving

$$\bar{I} = I_0 + \bar{I}_2. \tag{5.16}$$

We are interested in the wavelength shift, so only terms which lead to a positive change in one wing and a negative change in the other, i.e. which are antisymmetric with respect to the line centre, are of importance. Only the velocity-dependent terms have this property, and when combined with a (symmetric) temperature-dependent term they produce an antisymmetric product. The wavelength shift is then given approximately by

$$\overline{\Delta\lambda} = \bar{I}_{2a}/(\mathrm{d}I/\mathrm{d}\Delta\lambda), \tag{5.17}$$

where the antisymmetric part of the second-order perturbation is

$$\begin{aligned}
\bar{I}_{2a} = & \int_{-\infty}^{\infty} \left(\frac{\partial S}{\partial T} \frac{\partial \kappa}{\partial v_3} + S \frac{\partial^2 \kappa}{\partial T\, \partial v_3} \right) \mathrm{e}^{-\tau}\, \overline{T'(x_3)\, v_3'(x_3)}\, \mathrm{d}x_3 \\
& - \int_{-\infty}^{\infty} S\kappa\, \mathrm{e}^{-\tau} \int_{-\infty}^{x_3} \frac{\partial^2 \kappa}{\partial T\, \partial v_3} \overline{T'(x_3')\, v_3'(x_3')}\, \mathrm{d}x_3'\, \mathrm{d}x_3 \\
& - \int_{-\infty}^{\infty} \frac{\partial (S\kappa)}{\partial T} \mathrm{e}^{-\tau} \int_{-\infty}^{x_3} \frac{\partial \kappa}{\partial v_3} \overline{T'(x_3)\, v_3'(x_3')}\, \mathrm{d}x_3'\, \mathrm{d}x_3 \\
& - \int_{-\infty}^{\infty} S \frac{\partial \kappa}{\partial v_3} \mathrm{e}^{-\tau} \int_{-\infty}^{x_3} \frac{\partial \kappa}{\partial T} \overline{T'(x_3')\, v_3'(x_3)}\, \mathrm{d}x_3'\, \mathrm{d}x_3 \\
& + \int_{-\infty}^{\infty} S\kappa\, \mathrm{e}^{-\tau} \int_{-\infty}^{x_3} \frac{\partial \kappa}{\partial T} \int_{-\infty}^{x_3} \frac{\partial \kappa}{\partial v_3} \overline{T'(x_3')\, v_3'(x_3'')}\, \mathrm{d}x_3''\, \mathrm{d}x_3'\, \mathrm{d}x_3.
\end{aligned} \tag{5.18}$$

The line asymmetry is thus a linear function of the temperature derivatives of the opacity and the source function. Generally the opacity decreases and the source function increases with temperature. Thus the emission and the depth to which we see are increased; both effects contribute to produce greater intensities. If, at the same time, there is an upward velocity, the larger intensity gives the blueshift greater weight than the weaker redshifted profile. In the far wings of the line, where the continuous opacity in the neighbourhood of $\tau = 1$ contributes significantly, there is a complicated interplay between the relative abundances of the neutral hydrogen and the H^- ion. A temperature increase there tends to ionize the hydrogen, releasing electrons which then form H^- ions and *increase* the opacity. This moves the height of formation outwards to cooler regions and reduces the weight of the blueshifted profile and therefore the degree of asymmetry. The characteristic bow-like C-shape thus arises.

In the first two terms of Eqn. 5.18 the asymmetry is due to the mean value of the temperature and velocity product at each height. It thus reflects only the *correlated* portions of each field, in other words the convective flux directly,

$$\bar{F}_c(x_3) = \rho C_p \overline{T'(x_3)\, v_3'(x_3)}. \tag{5.19}$$

In the other terms, which contain a first-order fluctuation due to the changing opacity along the line of sight, the correlation between the fluctuations at *different* heights also plays a role. In principle, it would be possible to have an asymmetry without convective heat flux, if the thermodynamic quantities in different layers were correlated without being correlated in the same layer. At the centre of the disk, where the line of sight is vertical, such a picture is physically implausible: in the Sun, the C-shape is without doubt a result of the convective transport of heat. This is borne out by the calculations of Beckers & Nelson (1978), who used the semi-empirical model developed by Nelson & Musman (1977) to explore with qualitative success the convective origin of line shifts and asymmetries. In passing, we note that, since observation reveals no interdependence between the C-shape and the 5-min oscillations, we have direct evidence that the convective phenomena are indeed decoupled from acoustic oscillations, as we have assumed in Section 3.2.3.

Away from the centre of the disk, the two-point correlations become very important because the velocities influence the profile at a higher level in the atmosphere than the temperature fluctuations (see the two dotted lines in Fig. 5.11). When the inclined line of sight intersects a granule, as in case B, these

Fig. 5.11. Schematic model of the temperature and velocity field of the granulation. The shaded areas represent the hot granules. The mean levels at which temperature and velocity perturbations most influence the emergent intensities are indicated by the dotted lines T', v'. Various lines of sight are shown by the dashed lines A, A', etc. At large heliocentric angles C, C' the horizontal velocities contribute significantly to the velocity–brightness correlation and the net Fraunhofer line shifts.

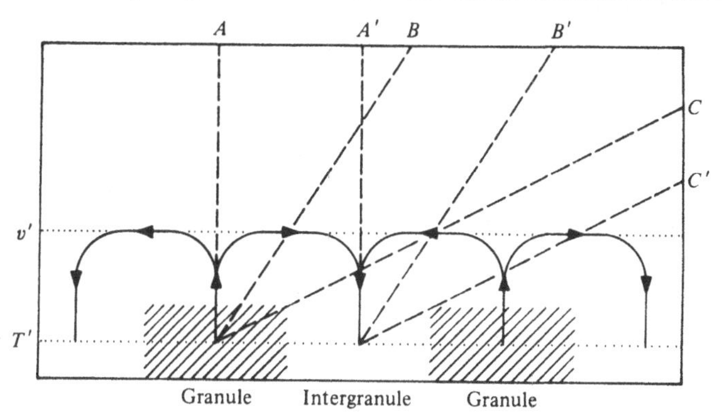

subsidiary terms correlate the horizontal flow towards the observer with the hot centre of the granule behind. The corresponding flow away from the observer is combined with the dark intergranular region (B'). Thus an additional 'convective' blueshift is contributed by the horizontal motions even in the absence of convective transport in the horizontal direction. Close to the limb, case C, the line of sight is further inclined and the separation between the points where the velocity and temperature contribute most to the correlation is correspondingly greater. Now the granular outflow towards the observer is combined with the dark intergranular lane lying behind the granule (C'). The left-hand granule combines with the outflow away from the observer in the granule in front of it (C). There is then a net redshift of the line, known sometimes as a 'supergravity' shift. The opposite signs of these horizontal contributions, together with the contribution of the vertical velocity, account for the departure of the limb shift curve (Fig. 5.8) from a simple cosine curve.

These features were qualitatively reproduced by the model of Beckers & Nelson but the quantitative comparison was not so satisfactory. The calculated shifts are about a factor of 2-3 too small, and the maximum occurs in the range $\cos\theta = 0.75$-0.6, with granular cell sizes of 1061-1556 km, instead of $\cos\theta \simeq 0.87$ as observed. The latter can be recovered by reducing the cell size and the former by increasing the amplitude of the temperature or velocity fluctuations (or both). Since the observed rms brightness fluctuation (12%) is reproduced by the model, Beckers & Nelson suggest that the measured velocity amplitudes used to calibrate the model are too small, especially in the region around $\tau = 0.3$. This would also bring their convective fluxes into better agreement with observation (see Fig. 5.6).

More realistic results have been achieved by the model of Dravins *et al.* (1981). These authors obtained a series of simulated mean profiles from part of the calculated time sequence described in Section 5.2.4. This portion excluded the strong exploding granule and the period of weak fluctuation following it. Thereafter the typical features of the observed C-shapes for lines of different strength were well reproduced, as may be seen in Fig. 5.9(c). These are full non-linear calculations and do not rely on the second-order weighting function described above. It may be remarked that these authors provide an explanation of the sharp reduction in the observed blueshift near the level of the continuum in terms of a greater velocity amplitude in the intergranular regions than in the granules, rather than in terms of a simple opacity effect.

The magnitude of the blueshift, of the order of 400 m s^{-1}, is correctly predicted. So the correct convective flux is present. This is achieved with larger temperature fluctuations than in the Beckers & Nelson model since the rms vertical velocity fluctiation (1.5 km s^{-1}) is about the same in both. The large temperature variations are equally apparent in the brightness variations; in Nordlund's model (upon which the analysis of Dravins *et al.* is based) the rms brightness is reported to lie between 20 and 30%. However, in the light of our discussion of brightness measurements (Section 2.5.2) these values seem significantly too large.

It is hard to avoid the conclusion that despite its success in qualitatively reproducing our picture of the granulation, Nordlund's simulation fails to deliver a quantitatively consistent synthesis of all its aspects. But the rapid strides being made in the computer simulation of such complicated physical systems holds great promise for refined modelling of the atmospheric convection pattern in the Sun and other stars.

5.6 Stellar granulation

Our discussion in the previous sections shows that there are two reliable avenues for detecting convective effects in other stars similar to those associated with the solar granulation. Characteristic signatures appear in the relative shifts of lines of different strength and in the asymmetries of individual lines. Both features have been advocated as stellar diagnostics, and attempts to exploit them are in progress.

The possibilities offered by line asymmetries were first explored by Bray & Loughhead (1978). They calculated synthetic integrated profiles for stars with a two-stream photospheric model and solar convective parameters. The stars were allowed to rotate with equatorial velocities of up to 30 km s^{-1}; Bray & Loughhead chose to demonstrate the asymmetry by means of the Fourier sine transform, which is identically zero for a symmetric profile but increases in amplitude with the degree of asymmetry. The greater the rotational broadening, the less obvious is the convective asymmetry. The amplitude of the sine transform is drastically reduced when the apparent equatorial speed exceeds 10 km s^{-1}. However, a solar-type rotation of approximately 2 km s^{-1} hardly influences the sine transform. There remains the problem of obtaining profiles of sufficient purity and freedom from blends. The latter difficulty is particularly troublesome as the sine transform requires the whole line profile to be measured and a blend anywhere within the line can fatally distort the results.

Faced with this problem, Gray (1980) attempted to isolate the asymmetry in only the central portions of 16 lines in Arcturus (spectral type K2 III). Instead of using Fourier transforms, he superposed the mirror image of the line on the original profile and matched parts of the upper wings. The difference between the two profiles in the core yields the asymmetry, the difference curve being characterized by its amplitude and width. The weaker lines showed no asymmetry; the stronger lines, in general, possessed cores shifted towards the blue with respect to the wings.

In a later investigation, Gray (1981) applied the same procedure to 28 lines in Procyon (spectral type F5 IV–V). In this star the red and blue wings differ, so the matching was not so straightforward. In all lines, the core is blueshifted with respect to the wings, with a modest increase in the magnitude of the shift with increasing line strength.

Gray (1982) has since obtained observations of higher spectral resolution using the coudé spectrograph of the 2.1-m telescope at the McDonald Observatory, Texas. From these he has derived line bisectors for 11 lines in the spectral region 623–626 nm in 27 F-, G- and K-type stars. Fig. 5.12(*a*) shows the line

Fig. 5.12. Examples of mean line bisectors in stars of various spectral types, divided according to luminosity class: (*a*) subgiants and dwarfs (IV–V); (*b*) giants (III). The Sun is included as a G2 V star. For clarity the error limits have been omitted; they range typically from $\pm 30\,\mathrm{m\,s^{-1}}$ in the middle of the bisector to $\pm 150\,\mathrm{m\,s^{-1}}$ near the continuum (Gray, 1982).

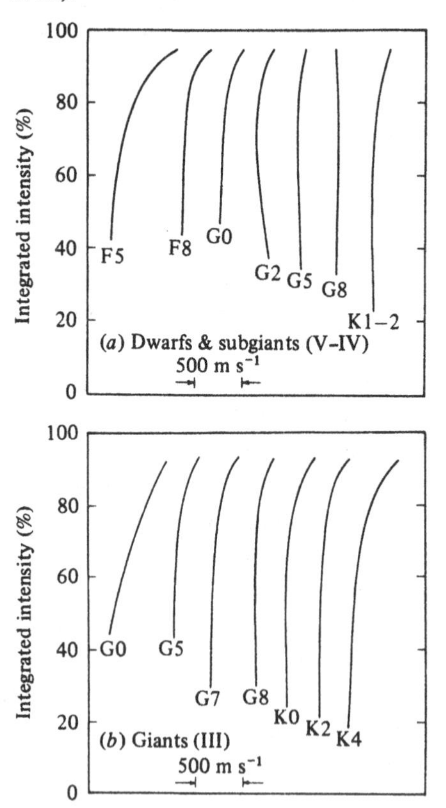

bisectors for dwarfs and subgiants (luminosity classes V and IV); note that those for G2 show typical solar behaviour. Fig. 5.12(*b*) shows the mean line bisectors for giants (luminosity class III). These tend to display a greater degree of asymmetry, with the wings showing a smaller blueward shift than the rest of the line. However, it is exceptional to also find a reduced blueshift in the line cores - the behaviour which gives the characteristic C-shape of solar-type stars. In main-sequence stars of late G-type the asymmetry almost vanishes. Gray did not measure the absolute wavelength position, and thus the magnitudes of the convective line shifts were not determined.

The approximate size of the wavelength shifts that would be expected on the basis of solar observations is $\sim 200\ \mathrm{m\ s^{-1}}$, an order of magnitude smaller than the Sun's equatorial rotational velocity. Any attempt to extract such shifts from observations of stellar lines requires spectral observations of high purity having an extremely accurate wavelength calibration. In Arcturus and Procyon the wavelengths determined photographically using a conventional spectrograph are subject to errors of about $75\ \mathrm{m\ s^{-1}}$ (Griffin & Griffin, 1973). Nevertheless, Dravins (1974) was able to show that the relative shifts of various lines in Arcturus are very similar to those in the Sun; on this basis he suggested that the convective velocities (more correctly, the convective fluxes) were similar in magnitude.

Fourier transform spectrographs automatically provide improved wavelength standards, and such interferometric instruments, used to record narrow segments of stellar spectra with extremely high spectral purity, are promising tools for extending the observations of convective phenomena in stars.

Existing theoretical models of stellar granulation are limited in scope. So far only Nelson (1980) has attempted to describe the lower atmosphere of a star - an F-type main-sequence star such as Procyon - by applying a solar-type model. The model is semi-empirical, and the exact values of horizontal scale and viscous drag length remain free parameters. At $\tau = 1$ the model yields vertical convective velocities of $1\text{-}2\ \mathrm{km\ s^{-1}}$ with a scale height of 150-1000 km and associated temperature fluctuations of 200-500 K, depending on the choice of parameters. The convective energy flux at the same depth is found to be less than 1% of the total flux, substantially lower than that in the Sun (Section 5.3). Nevertheless, the granular intensity pattern would have a higher contrast on an F-type star owing to the much more sensitive dependence of opacity on temperature. This study is but a first step, but an encouraging one in the sense that it predicts convective effects of at least the same magnitude as those in the Sun.

A theoretical understanding of the hydrodynamics of the upper convective layers of the Sun is within our grasp, but not of the convection zone as a whole. However, it would appear that the observed effects in stellar spectra can be accounted for entirely in terms of the outler layers. Furthermore, simulations of the photospheric and immediately subphotospheric layers are feasible with present-day computers, and they can be made self-consistent, i.e. they require no empirically determined parameters for calibration. This should provide the stimulus for the observational effort required to confirm theoretical models of convection in stars other than the Sun. Moreover, the recently identified global oscillations of the Sun (see Section 2.5.4), and perhaps of other stars, may offer the possibility of probing the *deeper* layers.

Today the treatment of convection in stars is rudimentary; but the tools, both theoretical and observational, now being developed encourage the belief that the next decade will witness the transformation of this branch of astrophysics into a field of rigorous investigation.

Additional notes

An improved treatment of turbulent cascade and mixing has been given by P. S. Marcus, W. H. Press & S. A. Teukolsky (1983, *Astrophysical Journal*, **267**, 795-821) (Sections 5.2.4 and 3.4.4).

The dynamical model of the supergranulation of R. van der Borght (1979, *Monthly Notices of the Royal Astronomical Society*, **188**, 615-24) was overlooked in Section 5.4. He attributes the supergranulation to convection driven by He II recombination.

The theory of the convective origin of the mean line profile asymmetry in Section 5.5.3 is described in more detail in M. Kaisig & C. J. Durrant (1982, *Astronomy and Astrophysics*, **116**, 332-40). It has been used by M. Kaisig & E. H. Schröter (1983, *Astronomy and Astrophysics*, **117**, 305-13) to investigate convection in active regions (Sections 5.5.2 and 5.5.3).

W. C. Livingston (1982, *Nature*, **297**, 208-9) claims to have detected a secular change in the global convective flux over the past five years (Sections 5.5.2 and 2.3.3).

REFERENCES

Chapter 1

Babcock, H. D. & Babcock, H. W. (1951). The ruling of diffraction gratings at the Mount Wilson Observatory. *Journal of the Optical Society of America*, **41**, 776–86.

Bartholomew, C. F. (1976). The discovery of the solar granulation. *Quarterly Journal of the Royal Astronomical Society*, **17**, 263–89.

Batchelor, G. K. (1953). *The Theory of Homogeneous Turbulence*. Cambridge University Press.

Burgers, J. M. & Thomas, R. N. (1958). Preface. In Proceedings of the Third Symposium on Cosmical Gas Dynamics. *Reviews of Modern Physics*, **30**, 908–10.

Chandrasekhar, S. (1949). Turbulence – a physical theory of astrophysical interest. *Astrophysical Journal*, **110**, 329-39.

Chevalier, S. (1908). Contribution to the study of the photosphere. *Astrophysical Journal*, **27**, 12-24.

Chevalier, S. (1914). Étude photographique de la photosphère solaire. *Annales de l'Observatoire Astronomique de Zô-Sè*, **8**, C1–24.

Dawes, W. R. (1864). Results of some recent observations of the solar surface, with remarks. *Monthly Notices of the Royal Astronomical Society*, **24**, 161–5.

Evans, J. W. (1963). Motions in the solar atmosphere. *Sky and Telescope*, **25**, 321–5.

Frenkiel, F. N. & Schwarzschild, M. (1955). Additional data for turbulence spectrum of solar photosphere at long wavelengths. *Astrophysical Journal*, **121**, 216–23.

Hansky, A. (1908). Mouvement des granules sur la surface du Soleil. *Mitteilungen des Astronomischen Hauptobservatoriums Pulkovo*, **3**, 1-20.

Herschel, W. (1801). Observations tending to investigate the nature of the Sun, in order to find the causes or symptoms of its variable emission of light and heat; with remarks on the use that may possibly be drawn from solar observations. *Philosophical Transactions of the Royal Society*, Part 1, p. 265.

Huggins, W. (1866). Results of some observations on the bright granules of the solar surface, with remarks on the nature of these bodies. *Monthly Notices of the Royal Astronomical Society*, **26**, 260-5.

Janssen, J. (1896). Mémoire sur la photographie solaire. *Annales de l'Observatoire d'Astronomie Physique de Paris sis Parc de Meudon (Seine-et-Oise)*, **1**, 91–102.

Kiepenheuer, K. O. (1953). Solar activity. In *The Sun*, ed. G. Kuiper, p. 322. University of Chicago Press.

King, H. C. (1955). *The History of the Telescope*. London: Griffin.

Langley, S. P. (1874). On the minute structure of the solar photosphere. *American Journal of Science*, Series 3, **7**, 87–101.

Langley, S. P. (1874). On the structure of the solar photosphere. *Monthly Notices of the Royal Astronomical Society*, **34**, 255–61.
Leighton, R. B. (1957). Some observations of solar granulation. *Publications of the Astronomical Society of the Pacific*, **69**, 497–505.
Loughhead, R. E. & Bray, R. J. (1959). 'Turbulence' and the photospheric granulation. *Nature*, **183**, 240–1.
Margrave, T. E. (1968). Review of visual observations of solar granulation. *Journal of the Washington Academy of Science*, **58**, No. 2, 26–31.
McMath, R. R., Mohler, O. C. & Pierce, A. K. (1955). Doppler shifts in solar granules. *Astrophysical Journal*, **122**, 565–6.
Nasmyth, J. (1862). On the structure of the luminous envelope of the Sun. *Memoirs of the Literary and Philosophical Society of Manchester*, Series 3, **1**, 407–11.
Plaskett, H. H. (1936). Solar granulation. *Monthly Notices of the Royal Astronomical Society*, **96**, 402–25.
Richardson, R. S. & Schwarzschild, M. (1950). On the turbulent velocities of solar granules. *Astrophysical Journal*, **111**, 351–61.
Rubashev, B. M. (1964). *Problems of Solar Activity*, Chapter 2. Moscow–Leningrad: Nauka Publishing House; Washington: NASA Technical Translation F-244.
Schwarzschild, M. & Schwarzschild, B. (1959). Balloon astronomy. *Scientific American*, **200**, No. 5, p. 52.
Secchi, A. (1875). *Le Soleil*, 2nd edn., vol. 1, Paris: Gauthier-Villars.
Siedentopf, H. (1933). Konvektion in Sternatmosphären. I. *Astronomische Nachrichten*, **247**, 297–306.
Skumanich, A. (1955). On bright–dark symmetry of solar granulation. *Astrophysical Journal*, **121**, 404–7.
Strebel, H. (1932). Sonnenphotographische Dokumente. *Zeitschrift für Astrophysik*, **5**, 36–49.
Strebel, H. (1933). Beitrag zum Problem der Sonnengranulation. *Zeitschrift für Astrophysik*, **6**, 313–29.
Uberoi, M. S. (1955). On the solar granules. *Astrophysical Journal*, **122**, 466–76.
Unsöld, A. (1930). Konvektion in der Sonnenatmosphäre. *Zeitschrift für Astrophysik*, **1**, 138–48.
Wasiutynski, J. (1946). Studies in hydrodynamics and structure of stars and planets. *Astrophysica Norvegica*, **4**, Chapter 4.
Young, C. A. (1895). *The Sun*. London: Kegan Paul.

Chapter 2

Aime, C., Martin, F., Grec, G. & Roddier, F. (1979). Statistical determination of a morphological parameter in solar granulation: spatial distribution of granules. *Astronomy and Astrophysics*, **79**, 1–7.
Albregtsen, F. & Hansen, T. L. (1977). The wavelength dependence of granulation (0.38–2.4 μm). *Solar Physics*, **54**, 31–3.
Alissandrakis, C. E., Macris, C. J. & Zachariadis, T. G. (1982). Measurements of the granule/intergranular lane contrast at 5200 Å and 6300 Å. *Solar Physics*, **76**, 129–36.
Allen, M. S. & Musman, S. (1973). The location of exploding granules. *Solar Physics*, **32**, 311–14.
Ando, H. & Osaki, Y. (1975). Nonadiabatic nonradial oscillations: an application to the five-minute oscillation of the Sun. *Publications of the Astronomical Society of Japan*, **27**, 581–603.

Bahng, J. & Schwarzschild, M. (1961). The temperature fluctuations in the solar granulation. *Astrophysical Journal*, **134**, 337–42.

Barletti, R., Ceppatelli, G., Paternò, L., Righini, A. & Speroni, N. (1977). Astronomical site testing with balloon-borne radiosondes: results about atmospheric turbulence, solar seeing and stellar scintillation. *Astronomy and Astrophysics*, **54**, 649–59.

Beckers, J. M. (1968a). High-resolution measurements of photosphere and sunspot velocity and magnetic fields using a narrow-band birefringent filter. *Solar Physics*, **3**, 258–68.

Beckers, J. M. (1968b). Photospheric brightness differences associated with the solar supergranulation. *Solar Physics*, **5**, 309–22.

Beckers, J. M. & Morrison, R. A. (1970). The interpretation of velocity filtergrams. III. Velocities inside solar granules. *Solar Physics*, **14**, 280–93.

Birkle, K. (1967). Über das Verhalten der photosphärischen Granulation im Fleckenzyklus. *Zeitschrift für Astrophysik*, **66**, 252–63.

Blackwell, D. E., Dewhirst, D. W. & Dollfus, A. (1959). The observation of solar granulation from a manned balloon. I. Observational data and measurement of contrast. *Monthly Notices of the Royal Astronomical Society*, **119**, 98–111.

Brandt, P. N. (1970). Measurement of solar image motion and blurring. *Solar Physics*, **13**, 243–6.

Brandt, P. N. (1976). *Messung der atmosphärisch bedingten Bildstörungen bei Sonnenbeobachtungen*. Doctoral Thesis, Freiburg: University of Freiburg.

Brandt, P. N. & Wöhl, H. (1982). Solar site-testing campaign of JOSO on the Canary Islands in 1979. *Astronomy and Astrophysics*, **109**, 77–89.

Bray, R. J. & Loughhead, R. E. (1958). Observations of changes in the photospheric granules. *Australian Journal of Physics*, **11**, 507–16.

Bray, R. J. & Loughhead, R. E. (1959). High resolution observations of the granular structure of sunspot umbrae. *Australian Journal of Physics*, **12**, 320–6.

Bray, R. J. & Loughhead, R. E. (1961). Facular granule lifetimes determined with a seeing-monitored photoheliograph. *Australian Journal of Physics*, **14**, 14–21.

Bray, R. J. & Loughhead, R. E. (1964). *Sunspots*. London: Chapman and Hall.

Bray, R. J. & Loughhead, R. E. (1967). *The Solar Granulation*. 1st edn. London: Chapman and Hall.

Bray, R. J. & Loughhead, R. E. (1974). *The Solar Chromosphere*. London: Chapman and Hall.

Bray, R. J. & Loughhead, R. E. (1977). A new determination of the granule/intergranule contrast. *Solar Physics*, **54**, 319–26.

Bray, R. J., Loughhead, R. E. & Norton, D. G. (1959). A 'seeing monitor' to aid solar observation. *The Observatory*, **79**, 63–5.

Bray, R. J., Loughhead, R. E. & Tappere, E. J. (1974). High-resolution photography of the solar chromosphere XV: preliminary observations in Fe I λ6569.2. *Solar Physics*, **39**, 323–6.

Bray, R. J., Loughhead, R. E. & Tappere, E. J. (1976). Convective velocities derived from granule contrast profiles in Fe I λ6569.2. *Solar Physics*, **49**, 3–18.

Canfield, R. C. (1976). The height variation of granular and oscillatory velocities. *Solar Physics*, **50**, 239–54.

Canfield, R. C. & Mehltretter, J. P. (1973). Fluctuations of brightness and vertical velocity at various heights in the photosphere. *Solar Physics*, **33**, 33–48.

Canfield, R. C. & Musman, S. (1973). Vertical phase variation and mechanical flux in the solar 5-minute oscillation. *Astrophysical Journal*, **184**, L131–6.

Carlier, A., Chauveau, F., Hugon, M. & Rösch, J. (1968). Cinématographie à haute résolution spatiale de la granulation photosphérique. *Comptes Rendus hebdomadaires des Séances de l'Académie des Sciences* B, **266**, 199–201.

Chandrasekhar, S. (1952). A statistical basis for the theory of stellar scintillation. *Monthly Notices of the Royal Astronomical Society*, **112**, 475–83.
Coulman, C. E. (1965). Optical image quality in a turbulent atmosphere. *Journal of the Optical Society of America*, **55**, 806–12.
Coulman, C. E. (1969). A quantitative treatment of solar 'seeing', I. *Solar Physics*, **7**, 122–43.
Cram, L. E. (1980). Solar physics. In *Advances in Electronics and Electron Physics*, vol. 54, ed. L. Marton & C. Marton, pp. 141–90. New York: Academic Press.
Dainty, J. C. (1975). Stellar speckle interferometry. In *Laser Speckle and Related Phenomena*, ed. J. C. Dainty, pp. 255–80. Berlin: Springer-Verlag.
Danielson, R. E. (1966). Observations of sunspot fine structure from the stratosphere and theoretical interpretation. In *Atti del Convegno sulle Macchie Solari*, ed. G. Righini, pp. 120–45. Firenze: G. Barbèra Editore.
Deubner, F. L. (1971). Some properties of velocity fields in the solar photosphere III: oscillatory and supergranular motions as a function of height. *Solar Physics*, **17**, 6–20.
Deubner, F. L. (1974a). On the energy distribution in wavenumber spectra of the granular velocity field. *Solar Physics*, **36**, 299–301.
Deubner, F. L. (1974b). Some properties of velocity fields in the solar photosphere V: spatio-temporal analysis of high resolution spectra. *Solar Physics*, **39**, 31–48.
Deubner, F. L. (1975). Observations of low wavenumber nonradial eigenmodes of the Sun. *Astronomy and Astrophysics*, **44**, 371–5.
Deubner, F. L. (ed.) (1977). The small scale structure of solar magnetic fields. In *Highlights of Astronomy*, vol. 4, part II, pp. 219–75. Dordrecht: D. Reidel Publishing Co.
Deubner, F. L. & Mattig, W. (1975). New observations of the granular intensity fluctuations. *Astronomy and Astrophysics*, **45**, 167–71.
Dravins, D., Lindegren, L. & Nordlund, Å. (1981). Solar granulation: influence of convection on spectral line asymmetries and wavelength shifts. *Astronomy and Astrophysics*, **96**, 345–64.
Dunn, R. B. (1981). *Solar Instrumentation: What's Next?* Sunspot: Sacramento Peak Observatory.
Dunn, R. B. & Zirker, J. B. (1973). The solar filigree. *Solar Physics*, **33**, 281–304.
Durrant, C. J., Kneer, F. & Maluck, G. (1981). The analysis of solar limb observations II. Geometrical smearing. *Astronomy and Astrophysics*, **104**, 211–14.
Durrant, C. J., Mattig, W., Nesis, A., Reiss, G. & Schmidt, W. (1979). Studies of granular velocities: VIII. The height dependence of the vertical granular velocity component. *Solar Physics*, **61**, 251–70.
Durrant, C. J. & Nesis, A. (1981). Vertical structure of the solar photosphere. *Astronomy and Astrophysics*, **95**, 221–8.
Duvall, T. L. (1980). The equatorial rotation rate of the supergranulation cells. *Solar Physics*, **66**, 213–21.
Edmonds, F. N. (1960). On solar granulation. *Astrophysical Journal*, **131**, 57–60.
Edmonds, F. N. (1962a). A coherence analysis of Fraunhofer line fine structure and continuum brightness fluctuations near the center of the solar disk. *Astrophysical Journal*, **136**, 507–33.
Edmonds, F. N. (1962b). A statistical photometric analysis of granulation across the solar disk. *Astrophysical Journal, Supplement Series*, **6**, 357–406.
Edmonds, F. N. & Hinkle, K. H. (1977). Spectral analyses of solar photospheric fluctuations V: A two-dimensional analysis of granulation at the centre of the disk. *Solar Physics*, **51**, 273–92.
Edmonds, F. N., Michard, R. & Servajean, R. (1965). Observational studies of macroscopic inhomogeneities in the solar atmosphere VII. A statistical analysis of photometric and kinematic inhomogeneities in the deep photosphere. *Annales d'Astrophysique*, **28**, 534–55.

Edmonds, F. N. & Webb, C. J. (1972a). Spectral analyses of solar photospheric fluctuations I. Power, coherence and phase spectra calculated by Fast-Fourier-Transform techniques. *Solar Physics*, **22**, 276-96.

Edmonds, F. N. & Webb, C. J. (1972b). Spectral analyses of solar photospheric fluctuations II. Bi-dimensional power, coherence and phase spectra of deep-seated radial velocity and photometric fluctuations. *Solar Physics*, **25**, 44-70.

Evans, J. W. & Michard, R. (1962a). Observational study of macroscopic inhomogeneities in the solar atmosphere II. Brightness fluctuations in Fraunhofer lines and the continuum. *Astrophysical Journal*, **136**, 487-92.

Evans, J. W. & Michard, R. (1962b). Observational study of macroscopic inhomogeneities in the solar atmosphere III. Vertical oscillatory motions in the solar photosphere. *Astrophysical Journal*, **136**, 493-506.

Frazier, E. N. (1968). A spatio-temporal analysis of velocity fields in the solar photosphere. *Zeitschrift für Astrophysik*, **68**, 345-56.

Frazier, E. N. (1970). Multi-channel magnetograph observations II: Supergranulation. *Solar Physics*, **14**, 89-111.

Fried, D. L. (1966). Optical resolution through a randomly inhomogeneous medium for very long and very short exposures. *Journal of the Optical Society of America*, **56**, 1372-9.

Fried, D. L. (ed.) (1977). Adaptive optics. In *Journal of the Optical Society of America*, **67**, 269-409.

Gaustad, J. & Schwarzschild, M. (1960). Note on the brightness fluctuation in the solar granulation. *Monthly Notices of the Royal Astronomical Society*, **121**, 260-2.

Giovanelli, R. G. & Slaughter, C. (1978). Motions in solar magnetic tubes I: The downflow. *Solar Physics*, **57**, 255-60.

Hagyard, M. J., West, E. A., Tandberg-Hanssen, E., Smith, J. E., Henze, W., Beckers, J. M., Bruner, E. C., Hyder, C. L., Gurman, J. B., Shine, R. A. & Woodgate, B. E. (1982). The photospheric vector magnetic field of a sunspot and its vertical gradient. In *The Physics of Sunspots*, ed. L. E. Cram & J. H. Thomas, pp. 213-34. Sunspot: Sacramento Peak Observatory.

Hart, A. B. (1954). Motions in the Sun at the photospheric level IV. The equatorial rotation and possible velocity fields in the photosphere. *Monthly Notices of the Royal Astronomical Society*, **114**, 17-38.

Hart, A. B. (1956). Motions in the Sun at the photospheric level VI. Large scale motions in the equatorial region. *Monthly Notices of the Royal Astronomical Society*, **116**, 38-55.

Harvey, J. W. & Ramsey, H. E. (1963). Photospheric granulation in the near ultraviolet. *Publications of the Astronomical Society of the Pacific*, **75**, 283-4.

Hejna, L. (1980). A comment on the character of the photospheric granular net. *Bulletin of the Astronomical Institutes of Czechoslovakia*, **31**, 362-4.

Howard, R. (1962). Preliminary solar magnetograph observations with small apertures. *Astrophysical Journal*, **136**, 211-22.

Hufnagel, R. E. & Stanley, N. R. (1964). Modulation transfer function associated with image transmission through turbulent media. *Journal of the Optical Society of America*, **54**, 52-61.

de Jager, C. (1959). Structure and dynamics of the solar atmosphere. In *Handbuch der Physik*, vol. 52, pp. 80-362. Berlin: Springer-Verlag.

Jenkins, G. M. & Watts, D. G. (1969). *Spectral Analysis and its Applications*. San Francisco: Holden-Day.

Joint Organization for Solar Observation (JOSO) (1970-9). *Annual Reports*. No publisher given.

Kallistratova, M. A. (1970). Measuring limb vibrations of solar images. In *Atmospheric Optics*, ed. N. B. Divari, pp. 9-17. New York: Consultants Bureau.

Karpinsky, V. N. (1979). Brightness and radial velocity are uncorrelated in the fine structure of the lower solar photosphere. *Soviet Astronomy Letters*, **5**, 295–7.

Karpinsky, V. N. (1980a). A morphological model of the fine structure of the photospheric brightness field. *Solnechnye Dannye Bjulleten*, **2**, 91–102.

Karpinsky, V. N. (1980b). Morphological elements and characteristics of fine structure of the photospheric brightness field near the solar disc center. *Solnechnye Dannye Bjulleten*, **7**, 94–103.

Karpinsky, V. N. & Mekhanikov, V. V. (1977). Two-dimensional spatial spectrum of the photospheric brightness field near to the solar disc center. *Solar Physics*, **54**, 25–30.

Karpinsky, V. N. & Pravdjuk, L. M. (1972). Variations of solar granulation with wavelength (from λ3900 Å to λ6600 Å). *Solnechnye Dannye Bjulleten*, **10**, 79–92.

Kawaguchi, I. (1980). Morphological study of the solar granulation II. The fragmentation of granules. *Solar Physics*, **65**, 207–20.

Keil, S. L. (1977). A new measurement of the center-to-limb variation of the rms granular contrast. *Solar Physics*, **53**, 359–68.

Keil, S. L. (1980). The structure of solar granulation I. Observations of the spatial and temporal behavior of vertical motions. *Astrophysical Journal*, **237**, 1024–34.

Keil, S. L. & Canfield, R. C. (1978). The height variation of velocity and temperature fluctuations in the solar photosphere. *Astronomy and Astrophysics*, **70**, 169–79.

Kirk, J. G. & Livingston, W. (1968). A solar granulation spectrogram. *Solar Physics*, **3**, 510–12.

Kitai, R. & Kawaguchi, I. (1979). Morphological study of the solar granulation I: Dark dot formation in the cell. *Solar Physics*, **64**, 3–12.

Kneer, F. (1973). On some characteristics of umbral fine structure. *Solar Physics*, **28**, 361–7.

Kneer, F., Mattig, W., Nesis, A. & Werner, W. (1980). Coherence analysis of granular intensity. *Solar Physics*, **68**, 31–9.

Korff, D., Dryden, G. & Miller, M. G. (1972). Information retrieval from atmospheric induced speckle patterns. *Optics Communications*, **5**, 187–92.

Koutchmy, S. & Adjabshirzadeh, A. (1981). Photometric analysis of the sunspot umbral dots: 2 - size, shape, and temperature. *Astronomy and Astrophysics*, **99**, 111–19.

Krat, V. A., Karpinsky, V. N. & Sobolev, V. M. (1972). Preliminary results of the third flight of the Soviet Stratospheric Solar Observatory. *Space Research*, **12**, vol. 2, 1713–17.

LaBonte, B. J., Simon, G. W. & Dunn, R. B. (1975). A phenomenological study of high-resolution granulation photographs. *Bulletin of the American Astronomical Society*, **7**, 366.

Leighton, R. B. (1963). The solar granulation. *Annual Review of Astronomy and Astrophysics*, **1**, 19–40.

Leighton, R. B., Noyes, R. W. & Simon, G. W. (1962). Velocity fields in the solar atmosphere. I. Preliminary report. *Astrophysical Journal*, **135**, 474–99.

Lévy, M. (1971). Analyse photométrique statistique de la granulation corrigée de l'influence de l'atmosphère terrestre et de l'instrument. *Astronomy and Astrophysics*, **14**, 15–23.

Livingston, W. C. (1968). Magnetograph observations of the quiet Sun. I. Spatial description of the background fields. *Astrophysical Journal*, **153**, 929–42.

Lohmann, A. W. & Weigelt, G. P. (1979). Astronomical speckle interferometry; measurements of isoplanicity and of temporal correlation. *Optik*, **53**, 167–80.

Loughhead, R. E. & Bray, R. J. (1960). Granulation near the extreme solar limb. *Australian Journal of Physics*, **13**, 738–9.

Loughhead, R. E. & Bray, R. J. (1961). Phenomena accompanying the birth of sunspot pores. *Australian Journal of Physics*, **14**, 347–51.

Loughhead, R. E. & Bray, R. J. (1966). Statistics of solar seeing. *Zeitschrift für Astrophysik*, **63**, 101–15.

Loughhead, R. E. & Bray, R. J. (1975). Visibility of the photospheric granulation in Fe I λ6569.2. *Solar Physics*, **45**, 35–40.
Loughhead, R. E., Bray, R. J. & Tappere, E. J. (1979). Improved observations of sunspot umbral dots. *Astronomy and Astrophysics*, **79**, 128–31.
Loughhead, R. E., Bray, R. J., Tappere, E. J. & Winter, J. G. (1968). High-resolution photography of the solar chromosphere I: the 30-cm refractor of the C.S.I.R.O. Solar Observatory. *Solar Physics*, **4**, 185–95.
Macris, C. J. (1953). Recherches sur la granulation photosphérique. *Annales d'Astrophysique*, **16**, 29–40.
Macris, C. (1959). The dimensions of the photospheric granules in various spectral regions. *The Observatory*, **79**, 22–4.
Macris, C. J. (1962). Studies on the flocculi of the solar chromosphere. Part I. Lifetime of the flocculi. *Memorie della Societa Astronomica Italiana*, **33**, 85–95.
Macris, C. J. (1978). The variation of the mean diameters of the photospheric granules near the sunspots. *Astronomy and Astrophysics*, **78**, 186–9.
Macris, C. J. & Banos, G. J. (1961). Mean distance between photospheric granules and its change with the solar activity. *Memoirs of the National Observatory Athens*, Series I, No. 8, 1–19.
Macris, C. & Elias, D. (1955). Sur une variation du nombre des granules photosphériques en fonction de l'activité solaire. *Annales d'Astrophysique*, **18**, 143–4.
Macris, C. & Prokakis, T. J. (1963). New results on the lifetime of the solar granules. *Memoirs of the National Observatory Athens*, Series I, No. 10, 1–10.
Maluck, G. (1980). *Bestimmung der Höhenabhängigkeit der granularen Intensitätsfluktuationen aus den Restintensitäten der Spektrostratoskopspektren.* Diplomarbeit. Freiburg: University of Freiburg.
Mattig, W., Mehltretter, J. P. & Nesis, A. (1981). Granular-size horizontal velocities in the solar atmosphere. *Astronomy and Astrophysics*, **96**, 96–101.
Mattig, W. & Nesis, A. (1974). Studies of granular velocities. IV: statistical analysis of granular Doppler-shifts. *Solar Physics*, **36**, 3–9.
Mehltretter, J. P. (1971a). Studies of granular velocities. II: Statistical analysis of two high-resolution spectrograms. *Solar Physics*, **16**, 253–71.
Mehltretter, J. P. (1971b). On the rms intensity fluctuation of solar granulation. *Solar Physics*, **19**, 32–9.
Mehltretter, J. P. (1973). Studies of granular velocities. III: the influence of finite spectral and spatial resolution upon the measurement of granular Doppler shifts. *Solar Physics*, **30**, 19–28.
Mehltretter, J. P. (1974). Observations of photospheric faculae at the center of the solar disk. *Solar Physics*, **38**, 43–57.
Mehltretter, J. P. (1978). Balloon-borne imagery of the solar granulation. II. The lifetime of solar granulation. *Astronomy and Astrophysics*, **62**, 311–16.
Mehltretter, J. P., Mattig, W. & von Alvensleben, A. (1978). *Projekt Spektrostratoskop. Sonnenforschung mit Stratosphärenballons.* Eggenstein: ZLDI. Forschungsbericht W78-15.
Meinel, A. B. (1960). Astronomical seeing and observatory site selection. In *Telescopes; Stars and Stellar Systems*, vol. 1, ed. G. P. Kuiper & B. M. Middlehurst, pp. 154–75. Chicago: University of Chicago Press.
Michard, R. (1961). Sur l'inhomogénéité de la photosphère solaire. *Comptes Rendus hebdomadaires des Séances de l'Académie des Sciences*, **252**, 4120–2.
Miller, R. A. (1960). Filamentary structure between sunspots photographed in integrated light. *Journal of the British Astronomical Association*, **70**, 100–1.
Moore, R. L. (1981). Dynamic phenomena in the visible layers of sunspots. *Space Science Reviews*, **28**, 387–421.

Muller, R. (1973). Etude morphologique et cinématique des structures fines d'une tache solaire. *Solar Physics*, **29**, 55-73.
Muller, R. (1976). The fine structure of photospheric sunspots and faculae. In *Solar Activity and Solar Terrestrial Relations*, ed. J. Sýkora, pp. 201-9. Bratislava: VEDA, Publishing House of the Slovak Academy of Science.
Muller, R. (1977). Morphological properties and origin of the photospheric facular granules. *Solar Physics*, **52**, 249-62.
Muller, R. (1981). Morphological and dynamic properties of magnetic bright points in the quiet photosphere. In *Proceedings of the Japan-France Seminar on Solar Physics*, ed. F. Moriyama & J. C. Henoux, pp. 142-8. No publisher given.
Musman, S. (1969). The effect of finite resolution on solar granulation. *Solar Physics*, **7**, 178-86.
Musman, S. (1974). The origin of the solar five-minute oscillation. *Solar Physics*, **36**, 313-19.
Namba, O. & Diemel, W. E. (1969). A morphological study of the solar granulation. *Solar Physics*, **7**, 167-77.
Namba, O. & van Rijsbergen, R. (1977). Evolution pattern of the exploding granules. In *Problems of Stellar Convection*, ed. E. A. Spiegel & J. P. Zahn, pp. 119-25. Berlin: Springer-Verlag.
November, L. J., Toomre, J., Gebbie, K. B. & Simon, G. W. (1981). The detection of meso-granulation on the Sun. *Astrophysical Journal*, **245**, L123-L126.
Plaskett, H. H. (1955). Physical conditions in the solar photosphere. In *Vistas in Astronomy*, vol. 1, ed. A. Beer, pp. 637-47. London: Pergamon Press.
Pravdjuk, L. M., Karpinsky, V. N. & Andreiko, A. V. (1974). A distribution of the brightness amplitudes in photospheric granulation. *Solnechnye Dannye Bjulleten*, **2**, 70-88.
Priestley, C. H. B. (1959). *Turbulent Transfer in the Lower Atmosphere*. Chicago: University of Chicago Press.
Rhodes, E. J., Ulrich, R. K. &Simon, G. W. (1977). Observations of non-radial p-mode oscillations on the Sun. *Astrophysical Journal*, **218**, 901-19.
Richardson, R. S. & Schwarzschild, M. (1950). On the turbulent velocities of solar granules. *Astrophysical Journal*, **111**, 351-61.
Ricort, G. & Aime, C. (1979). Solar seeing and the statistical properties of the photospheric solar granulation. III. Solar speckle interferometry. *Astronomy and Astrophysics*, **76**, 324-35.
Ricort, G., Aime, C., Deubner, F. L. & Mattig, W. (1981). Solar granulation study in partial eclipse conditions using speckle interferometric techniques. *Astronomy and Astrophysics*, **97**, 114-21.
Roddier, F. (1981). The effects of atmospheric turbulence in optical astronomy. In *Progress in Optics*, vol. 19, ed. E. Wolf, pp. 283-376. Amsterdam: North-Holland Publishing Company.
Rösch, J. (1957). Photographies de la photosphère et des taches solaires. *L'Astronomie*, **71**, 129-41.
Rösch, J. (1959). Observations sur la photosphère solaire. II. Numération et photométrie photographique des granules dans le domaine spectral 5900-6000 Å. *Annales d'Astrophysique*, **22**, 584-607.
Rösch, J. (1962). Results drawn from photographs of the photosphere obtained from the ground. In *Transactions of the International Astronomical Union*, vol. 11B, ed. D. H. Sadler, p. 197. London: Academic Press.
Rösch, J. (ed.) (1963). La choix des sites d'observatoire astronomique (site testing) - IAU Symposium 19. *Bulletin Astronomique*, **24**, 313.
Rösch, J. & Hugon, M. (1959). Sur l'évolution dans le temps de la granulation photosphérique. *Comptes Rendus hebdomadaires des Séances de l'Académie des Sciences*, **249**, 625-7.

Schmidt, W., Deubner, F. L., Mattig, W. & Mehltretter, J. P. (1979). On the center to limb variation of the granular brightness fluctuations. *Astronomy and Astrophysics*, **75**, 223-7.

Schmidt, W., Knölker, M. & Schröter, E. H. (1981). Rms-value and power spectrum of the photospheric intensity fluctuations. *Solar Physics*, **73**, 217-31.

Schröter, E. H. (1957). Zur Deutung der Rotverschiebung und der Mitte-Rand-Variation der Fraunhoferlinien bei Berücksichtigung der Temperaturschwankungen der Sonnenatmosphäre. *Zeitschrift für Astrophysik*, **41**, 141-81.

Schröter, E. H. (1962). Einige Beobachtungen und Messungen an Stratoskop I-Negativen. *Zeitschrift für Astrophysik*, **56**, 183-93.

Schwarzschild, M. (1959). Photographs of the solar granulation taken from the stratosphere. *Astrophysical Journal*, **130**, 345-63.

Semel, M. (1962). Sur la détermination du champ magnétique de la granulation solaire. *Comptes Rendus hebdomadaires des Séances de l'Académie des Sciences*, **254**, 3978-80.

Simon, G. W. (1964). Calcium network and vertical velocities. In *Transactions of the International Astronomical Union*, vol. 12B, ed. J. C. Pecker, p. 164. London: Academic Press.

Simon, G. W. (1967). Observations of horizontal motions in solar granulation: their relation to supergranulation. *Zeitschrift für Astrophysik*, **65**, 345-63.

Simon, G. W. & Leighton, R. B. (1964). Velocity fields in the solar atmosphere. III. Large-scale motions, the chromospheric network, and magnetic fields. *Astrophysical Journal*, **140**, 1120-47.

Simon, G. W. & Zirker, J. B. (1974). A search for the footpoints of solar magnetic fields. *Solar Physics*, **35**, 331-42.

Sobolev, V. M. (1975). On the profile characteristics of the Fraunhofer lines in the granules and intergranular regions. *Solnechnye Dannye Bjulleten*, **6**, 68-71.

Steshenko, N. V. (1960). On the determination of magnetic fields of solar granulation. *Izvestiya Krymskoj Astrofizicheskoj Observatorii*, **22**, 49-55.

Stock, J. & Keller, G. (1960). Astronomical seeing. In *Telescopes; Stars and Stellar Systems*, vol. 1, ed. G. P. Kuiper & B. M. Middlehurst, pp. 138-53. Chicago: University of Chicago Press.

Strebel, H. (1932). Sonnenphotographische Dokumente. *Zeitschrift für Astrophysik*, **5**, 36-49.

Strebel, H. (1933). Beitrag zum Problem der Sonnengranulation. *Zeitschrift für Astrophysik*, **6**, 313-29.

Stuart, F. E. & Rush, J. H. (1954). Correlation analyses of turbulent velocities and brightness of the photospheric granulation. *Astrophysical Journal*, **120**, 245-50.

Tarbell, T. D. & Smithson, R. C. (1981). A simple image motion compensation system for solar observations. In *Solar Instrumentation: What's Next?*, ed. R. B. Dunn, pp. 491-501. Sunspot: Sacramento Peak Observatory.

Tatarski, V. I. (1961). *Wave Propagation in a Turbulent Medium*. New York: McGraw-Hill (republished by Dover Publications Inc. 1967).

Uberoi, M. G. & Kovasznay, L. S. G. (1953). On mapping and measurement of random fields. *Quarterly Journal of Applied Mathematics*, **10**, 375-93.

Ulrich, R. K. (1970). The five-minute oscillations on the solar surface. *Astrophysical Journal*, **162**, 993-1002.

Webb, E. K. (1964). Daytime thermal fluctuations in the lower atmosphere. *Applied Optics*, **3**, 1329-36.

Weigelt, G. P. (1978). Speckle holography measurements of the stars Zeta Cancri and ADS 3358. *Applied Optics*, **17**, 2660-2.

Wiesmeier, A. & Durrant, C. J. (1981). The analysis of solar limb observations. I. Restoration of data in a tilted reference frame. *Astronomy and Astrophysics*, **104**, 207-10.

Wittmann, A. (1981). Balloon-borne imagery of the solar granulation. III. Digital analysis of a white-light time series. *Astronomy and Astrophysics*, **99**, 90–6.
Wittmann, A. & Mehltretter, J. P. (1977). Balloon-borne imagery of the solar granulation. I. Digital image enhancement and photometric properties. *Astronomy and Astrophysics*, **61**, 75–8.
Worden, S. P. (1975). Infrared observations of supergranule temperature structure. *Solar Physics*, **45**, 521–32.
Worden, S. P. & Simon, G. W. (1976). A study of supergranulation using a diode array magnetograph. *Solar Physics*, **46**, 73–91.
Young, A. T. (1974). Seeing: its cause and cure. *Astrophysical Journal*, **189**, 587–604.

Chapter 3

Adrian, R. J. (1975). Turbulent convection in water over ice. *Journal of Fluid Mechanics*, **69**, 753–81.
Ahlers, G. & Behringer, R. P. (1978). Evolution of turbulence from the Rayleigh-Bénard instability. *Physical Review Letters*, **40**, 712–16.
Block, M. J. (1956). Surface tension as the cause of Bénard cells and surface deformation in a liquid film. *Nature*, **178**, 650–1.
Böhm, K. H. & Richter, E. (1959). Der Einfluß der Strahlungsaustausches auf die Konvektion in einer polytropen Atmosphäre. *Zeitschrift für Astrophysik*, **48**, 231–48.
Böhm, K. H. & Richter, E. (1960). Konvektion in einer Atmosphäre mit tiefenabhängigem Temperaturgradienten und starker Dichtevariation. *Zeitschrift für Astrophysik*, **50**, 79–95.
Bradshaw, P. (ed.) (1976). *Turbulence*. Berlin: Springer-Verlag.
Bray, R. J. & Loughhead, R. E. (1974). *The Solar Chromosphere*. London: Chapman and Hall.
Busse, F. H. (1967). On the stability of two-dimensional convection in a layer heated from below. *Journal of Mathematics and Physics*, **46**, 140–50.
Busse, F. H. (1972). The oscillatory instability of convection rolls in a low Prandtl number fluid. *Journal of Fluid Mechanics*, **52**, 97–112.
Busse, F. H. & Clever, R. M. (1979). Instabilities of convection rolls in a fluid of moderate Prandtl number. *Journal of Fluid Mechanics*, **91**, 319–35.
Busse, F. H. & Whitehead, J. A. (1971). Instabilities of convection rolls in a high Prandtl number fluid. *Journal of Fluid Mechanics*, **47**, 305–20.
Busse, F. H. & Whitehead, J. A. (1974). Oscillatory and collective instabilities in large Prandtl number convection. *Journal of Fluid Mechanics*, **66**, 67–79.
Chandrasekhar, S. (1939). *An Introduction to the Study of Stellar Structure*. Chicago: University of Chicago Press (republished by Dover Publications, Inc., 1957).
Chandrasekhar, S. (1961). *Hydrodynamic and Hydromagnetic Stability*. Oxford: Oxford University Press (republished by Dover Publications, Inc. 1981).
Clever, R. M. & Busse, F. H. (1974). Transition to time-dependent convection. *Journal of Fluid Mechanics*, **65**, 625–45.
Cox, J. P. & Giuli, R. T. (1968). *Principles of Stellar Structure. Volume 1: Physical Principles*. New York: Gordon and Breach.
Daniels, P. G. (1977). The effect of distant sidewalls on the transition to finite amplitude Bénard convection. *Proceedings of the Royal Society of London*, Series A, **358**, 173–97.
Goldstein, R. J. & Graham, D. J. (1969). Stability of a horizontal fluid layer with zero shear boundaries. *Physics of Fluids*, **12**, 1133–7.
Gough, D. O. (1969). The anelastic approximation for thermal convection. *Journal of the Atmospheric Sciences*, **26**, 448–56.

Gough, D. O., Moore, D. R., Spiegel, E. A. & Weiss, N. O. (1976). Convective instability in a compressible atmosphere. II. *Astrophysical Journal*, **206**, 536–42.
Gough, D. O., Spiegel, E. A. & Toomre, J. (1975). Modal equations for cellular convection. *Journal of Fluid Mechanics*, **68**, 695–719.
Graham, E. & Moore, D. R. (1978). The onset of compressible convection. *Monthly Notices of the Royal Astronomical Society*, **183**, 617–32.
Grodzka, P. G. & Bannister, T. C. (1975). Heat flow and convection experiments aboard Apollo 17. *Science*, **187**, 165–7.
Hall, P. & Walton, I. C. (1977). The smooth transition to a convective regime in a two-dimensional box. *Proceedings of the Royal Society of London*, Series A, **358**, 199–221.
Hall, P. & Walton, I. C. (1979). Bénard convection in a finite box: secondary and imperfect bifurcations. *Journal of Fluid Mechanics*, **90**, 377–95.
Jones, C. A. & Moore, D. R. (1979). The stability of axisymmetric convection. *Geophysical and Astrophysical Fluid Dynamics*, **11**, 245–70.
Jones, C. A., Moore, D. R. & Weiss, N. O. (1976). Axisymmetric convection in a cylinder. *Journal of Fluid Mechanics*, **73**, 353–88.
Koschmieder, E. L. (1974). Bénard convection. In *Advances in Chemical Physics*, vol. 26, ed. I. Prigogine & S. A. Rice, pp. 177–212. New York: John Wiley & Sons.
Koschmieder, E. L. & Pallas, S. G. (1974). Heat transfer through a shallow, horizontal convecting fluid layer. *International Journal of Heat and Mass Transfer*, **17**, 991–1002.
Krishnamurti, R. (1970a). On the transition to turbulent convection. Part 1. The transition from two- to three-dimensional flow. *Journal of Fluid Mechanics*, **42**, 295–307.
Krishnamurti, R. (1970b). On the transition to turbulent convection. Part 2. The transition to time-dependent flow. *Journal of Fluid Mechanics*, **42**, 309–20.
Krishnamurti, R. (1973). Some further studies on the transition to turbulent convection. *Journal of Fluid Mechanics*, **60**, 285–303.
Landau, L. D. & Lifshitz, E. M. (1959). *Fluid Mechanics*. Oxford: Pergamon Press.
Liepmann, H. W. (1979). The rise and fall of ideas in turbulence. *American Scientist*, **67**, 221–8.
Malkus, W. V. R. (1954). Discrete transitions in turbulent convection. *Proceedings of the Royal Society of London*, Series A, **225**, 185–95.
Malkus, W. V. R. & Veronis, G. (1958). Finite amplitude cellular convection. *Journal of Fluid Mechanics*, **4**, 225–60.
Mihalas, B. W. & Toomre, J. (1981). Internal gravity waves in the solar atmosphere. I. Adiabatic waves in the chromosphere. *Astrophysical Journal*, **249**, 349–71.
Mihalas, B. W. & Toomre, J. (1982). Internal gravity waves in the solar atmosphere. II. Effects of radiative damping. *Astrophysical Journal*, **263**, 386–408.
Moffatt, H. K. (1981). Some developments in the theory of turbulence. *Journal of Fluid Mechanics*, **106**, 27–47.
Moore, D. R. & Weiss, N. O. (1973a). Two-dimensional Rayleigh–Bénard convection. *Journal of Fluid Mechanics*, **58**, 289–312.
Moore, D. R. & Weiss, N. O. (1973b). Nonlinear penetrative convection. *Journal of Fluid Mechanics*, **61**, 553–81.
Moore, D. W. (1967). Prediction on the velocity field coming from convective overshoot in a layer overlying a convectively unstable atmospheric region. In *Aerodynamic Phenomena in Stellar Atmospheres*, IAU Symposium 28, ed. R. N. Thomas, pp. 405–14. London: Academic Press.
Morse, P. M. (1965). *Thermal Physics*. New York: Benjamin.
Murphy, J. O. (1971). Non-linear convection with free boundaries at high Rayleigh numbers. *Proceedings of the Astronomical Society of Australia*, **2**, 51–2.
Musman, S. (1968). Penetrative convection. *Journal of Fluid Mechanics*, **31**, 343–60.

Myrup, L., Gross, D., Hoo, L. S. & Goddard, W. (1970). Upside down convection. *Weather*, **25**, 150-7.

Normand, C., Pomeau, Y. & Valverde, M. G. (1977). Convective instability: A physicist's approach. *Reviews of Modern Physics*, **49**, 581-624.

Palm, E. (1960). On the tendency towards hexagonal cells in steady convection. *Journal of Fluid Mechanics*, **8**, 183-92.

Palm, E. (1975). Nonlinear thermal convection. *Annual Review of Fluid Mechanics*, **7**, 39-61.

Pearson, J. R. A. (1958). On convection cells induced by surface tension. *Journal of Fluid Mechanics*, **4**, 489-500.

Rayleigh, Lord (1916). On convection currents in a horizontal layer of fluid when the higher temperature is on the under side. *Philosophical Magazine*, Series 6, **32**, 529-46.

Schlüter, A., Lortz, D. & Busse, F. H. (1965). On the stability of steady finite amplitude convection. *Journal of Fluid Mechanics*, **23**, 129-44.

Schwarzschild, K. (1906). Über das Gleichgewicht der Sonnenatmosphäre. *Nachrichten der Königlichen Gesellschaft der Wissenschaften zu Göttingen*, 41 (English translation in: Meadows, A. J. (1970). *Early Solar Physics*. Oxford: Pergamon Press.)

Schwarzschild, M. (1958). *Structure and Evolution of the Stars*. Chapter 2. Princeton University Press.

Segel, L. A. & Stuart, J. T. (1962). On the question of the preferred mode in cellular thermal convection. *Journal of Fluid Mechanics*, **13**, 289-306.

Skumanich, A. (1955). On thermal convection in a polytropic atmosphere. *Astrophysical Journal*, **121**, 408-17.

Spiegel, E. A. (1964). The effect of radiative transfer on convective growth rates. *Astrophysical Journal*, **139**, 959-74.

Spiegel, E. A. (1965). Convective instability in a compressible atmosphere. I. *Astrophysical Journal*, **141**, 1068-90.

Stix, M. (1970). Two examples of penetrative convection. *Tellus*, XXII, **5**, 517-20.

Stuart, J. T. (1964). On the cellular patterns in thermal convection. *Journal of Fluid Mechanics*, **18**, 481-98.

Tennekes, H. & Lumley, J. L. (1972). *A First Course in Turbulence*. Cambridge, Mass.: MIT Press.

Townsend, A. A. (1964). Natural convection in water over an ice surface. *Quarterly Journal of the Royal Meteorological Society*, **90**, 248-59.

Townsend, A. A. (1966). Internal waves produced by a convective layer. *Journal of Fluid Mechanics*, **24**, 307-19.

Truesdell, C. & Muncaster, R. G. (1980). *Fundamentals of Maxwell's Kinetic Theory of a Simple Monatomic Gas*. New York: Academic Press.

Unno, W., Kato, S. & Makita, M. (1960). Convective instability in polytropic atmospheres. I. *Publications of the Astronomical Society of Japan*, **12**, 192-202.

Vardya, M. S. (1965). Thermodynamics of a solar composition gaseous mixture. *Monthly Notices of the Royal Astronomical Society*, **129**, 205-13.

Veronis, G. (1963). Penetrative convection. *Astrophysical Journal*, **137**, 641-63.

Vickers, G. T. (1971). On the formation of giant cells and supergranules. *Astrophysical Journal*, **163**, 363-74.

Weiss, N. O. (1977). Numerical methods in convection theory. In *Problems of Stellar Convection*, eds. E. A. Spiegel & J. P. Zahn, pp. 142-50. Berlin: Springer-Verlag.

Whitehead, J. A. (1971). Cellular convection. *American Scientist*, **59**, 444-51.

Willis, G. E. & Dearsdorff, J. W. (1967). Confirmation and renumbering of the discrete heat flux transitions of Malkus. *Physics of Fluids*, **10**, 1861-7.

Willis, G. E. & Dearsdorff, J. W. (1970). The oscillatory motions of Rayleigh convection. *Journal of Fluid Mechanics*, **44**, 661-72.

Yih, C.-S. (1969). *Fluid Mechanics. A Concise Introduction to the Theory*. New York: McGraw-Hill.

Chapter 4

Athay, R. G. & Lites, B. W. (1972). Fe I ionization and excitation equilibrium in the solar atmosphere. *Astrophysical Journal*, **176**, 809-31.

Beckers, J. M. & Milkey, R. W. (1975). The line response function of stellar atmospheres and the effective depth of line formation. *Solar Physics*, **43**, 289-92.

Böhm-Vitense, E. (1958). Über die Wasserstoffkonvektionszone in Sternen verschiedener Effektivetemperaturen und Leuchtkräfte. *Zeitschrift für Astrophysik*, **46**, 108-43.

Böhm, K. H. (1963a). Unstable modes in the solar hydrogen convection zone. *Astrophysical Journal*, **137**, 881-900.

Böhm, K. H. (1963b). Strömungsformen verschiedener vertikaler Wellenlängen in der solaren Wasserstoffkonvektionszone. *Zeitschrift für Astrophysik*, **57**, 265-77.

Cannon, C. J. (1970). Line transfer in two dimensions. *Astrophysical Journal*, **161**, 255-64.

Cannon, C. J. (1976). Solutions of the radiative transfer equation for Lyα, Lyβ, Mg II h and k, Ca II H and K using two-dimensional geometry, macroscopic velocity fields, and frequency and angle dependent redistribution. *Astronomy and Astrophysics*, **52**, 337-62.

Cox, J. P. & Giuli, R. T. (1968). *Principles of Stellar Structure. Volume 1: Physical Principles*. New York: Gordon and Breach.

Deupree, R. G. (1975). On shallow convective envelopes. *Astrophysical Journal*, **201**, 183-9.

Deupree, R. G. (1976). Nonlinear convective motion in shallow convective envelopes. *Astrophysical Journal*, **205**, 286-94.

Durney, B. R. & Spruit, H. C. (1979). On the dynamics of stellar convection zones: The effect of rotation on the turbulent viscosity and conductivity. *Astrophysical Journal*, **234**, 1067-78.

Edmonds, F. N. (1957). The coefficients of viscosity and thermal conductivity in the hydrogen convection zone. *Astrophysical Journal*, **125**, 535-49.

Eichler, D. (1977). A spectral model of turbulent convection. *Astrophysical Journal*, **211**, 894-99.

Eschrich, K. O. (1978). Wärmetransport und Wärmeerzeugung in einem turbulent bewegten Gas. *Astronomische Nachrichten*, **299**, 137-44.

Giovanelli, R. G. (1959). Radiative transfer in non-uniform media. *Australian Journal of Physics*, **12**, 164-70.

Gough, D. O. (1969). The anelastic approximation for thermal convection. *Journal of the Atmospheric Sciences*, **26**, 448-56.

Gough, D. O. (1977a). The current state of stellar mixing-length theory. In *Problems of Stellar Convection*, eds. E. A. Spiegel & J. P. Zahn, pp. 15-56. Berlin: Springer-Verlag.

Gough, D. O. (1977b). Stellar convection. In *Problems of Stellar Convection*, eds. E. A. Spiegel & J. P. Zahn, pp. 349-63. Berlin: Springer-Verlag.

Gough, D. O., Moore, D. R., Spiegel, E. A. & Weiss, N. O. (1976). Convective instability in a compressible atmosphere. II. *Astrophysical Journal*, **206**, 536-42.

Gough, D. O. & Weiss, N. O. (1976). The calibration of stellar convection theories. *Monthly Notices of the Royal Astronomical Society*, **176**, 589-607.

Gouttebroze, P. & Leibacher, J. W. (1980). Solar atmospheric dynamics. I. Formation of optically thick chromospheric lines. *Astrophysical Journal*, **238**, 1134-51.

Graham, E. (1975). Numerical simulation of two-dimensional compressible convection. *Journal of Fluid Mechanics*, **70**, 689-703.

Graham, E. (1977). Compressible convection. In *Problems of Stellar Convection*, eds. E. A. Spiegel & J. P. Zahn, pp. 151-5. Berlin: Springer-Verlag.

Hart, M. H. (1973). Linear convective models and the energy transport in stellar convection zones. *Astrophysical Journal*, **184**, 587–603.

Howard, R. & LaBonte, B. J. (1980). A search for large-scale convection cells in the solar atmosphere. *Astrophysical Journal*, **239**, 738–45.

Jones, H. P. & Skumanich, A. (1980). The physical effects of radiative transfer in multidimensional media including models of the solar atmosphere. *Astrophysical Journal, Supplement Series*, **42**, 221–40.

Keil, S. L. (1980). The interpretation of solar line shift observations. *Astronomy and Astrophysics*, **82**, 144–51.

Kneer, F. (1980). Multidimensional radiative transfer in stratified atmospheres. III. Non-LTE line formation. *Astronomy and Astrophysics*, **93**, 387–94.

Kneer, F. & Heasley, J. N. (1979). Multidimensional radiative transfer in stratified atmospheres: gray radiative equilibrium. *Astronomy and Astrophysics*, **79**, 14–21.

Knobloch, E. (1977). The diffusion of scalar and vector fields by homogeneous stationary turbulence. *Journal of Fluid Mechanics*, **83**, 129–40.

Krause, F. & Rüdiger, G. (1974). On the Reynolds stresses in mean-field hydrodynamics. I. Incompressible homogeneous isotropic turbulence. *Astronomische Nachrichten*, **295**, 93–9.

Latour, J., Spiegel, E. A., Toomre, J. & Zahn, J. P. (1976). Stellar convection theory. I. The anelastic modal equations. *Astrophysical Journal*, **207**, 233–43.

Ledoux, P., Schwarzschild, M. & Spiegel, E. A. (1961). On the spectrum of turbulent convection. *Astrophysical Journal*, **133**, 184–97.

Lévy, M. (1974). Temperature fluctuations in solar granulation. *Astronomy and Astrophysics*, **31**, 451–8.

Lorenz, E. N. (1963). Deterministic nonperiodic flow. *Journal of the Atmospheric Sciences*, **20**, 130–41.

Massaguer, J. M. & Zahn, J. P. (1980). Cellular convection in a stratified atmosphere. *Astronomy and Astrophysics*, **87**, 315–27.

Mihalas, D. (1978). *Stellar Atmospheres.* 2nd edition. San Francisco: W. H. Freeman.

Moore, D. R. (1979). Can granulation be understood as convective overshoot? In *Small Scale Motions on the Sun*, pp. 55–61. Freiburg: Kiepenheuer-Institut.

Moore, D. W. (1967). Prediction on the velocity field coming from convective overshoot in a layer overlying a convectively unstable atmospheric region. In *Aerodynamic Phenomena in Stellar Atmospheres* (IAU Symposium 28), ed. R. N. Thomas, pp. 405–14. London: Academic Press.

Mullan, D. S. (1971). Cellular convection in model stellar envelopes. *Monthly Notices of the Royal Astronomical Society*, **154**, 467–89.

Nordlund, Å. (1974). On convection in stellar atmospheres. *Astronomy and Astrophysics*, **32**, 407–22.

Nordlund, Å. (1976). A two-component representation of stellar atmospheres with convection. *Astronomy and Astrophysics*, **50**, 23–39.

Öpik, E. J. (1950). Transport of heat and matter by convection in stars. *Monthly Notices of the Royal Astronomical Society*, **110**, 559–89.

Owocki, S. P. & Auer, L. H. (1980). Two dimensional radiative transfer. II. The wings of Ca K and Mg k. *Astrophysical Journal*, **241**, 448–58.

Parsons, S. B. (1969). Model atmospheres for yellow supergiants. *Astrophysical Journal, Supplement Series*, **18**, 127–65.

Roxburgh, I. W. & Tavakol, R. K. (1979). The origin of supergranulation and giant cells in the solar convection zone. *Solar Physics*, **61**, 247–50.

Spiegel, E. A. (1957). The smoothing of temperature fluctuations by radiative transfer. *Astrophysical Journal*, **126**, 202–7.

Spiegel, E. A. (1963). A generalization of the mixing-length theory of turbulent convection. *Astrophysical Journal*, **138**, 216–25.
Spiegel, E. A. (1971). Convection in stars: I. Basic Boussinesq convection. *Annual Review of Astronomy and Astrophysics*, **9**, 323–52.
Spiegel, E. A. (1972). Convection in stars: II. Special effects. *Annual Review of Astronomy and Astrophysics*, **10**, 261–304.
Spruit, H. C. (1974). A model of the solar convection zone. *Solar Physics*, **34**, 277–90.
Swinney, H. L. & Gollub, J. P. (eds.) (1981). *Hydrodynamic Instabilities and the Transition to Turbulence*. Berlin: Springer-Verlag.
Toomre, J., Zahn, J. P., Latour, J. & Spiegel, E. A. (1976). Stellar convection theory. II. Single-mode study of the second convection zone in an A-type star. *Astrophysical Journal*, **207**, 545–63.
Travis, L. D. & Matsushima, S. (1973). The role of convection in stellar atmospheres. I. Observable effects of convection in the solar atmosphere. *Astrophysical Journal*, **180**, 975–85.
Turner, J. S. (1964). The flow into an expanding spherical vortex. *Journal of Fluid Mechanics*, **18**, 195–208.
Ulrich, R. K. (1970a). Convective energy transport in stellar atmospheres. I: A convective thermal model. *Astrophysics and Space Science*, **7**, 71–86.
Ulrich, R. K. (1970b). Convective energy transport in stellar atmospheres. II: Model atmosphere calculations. *Astrophysics and Space Science*, **7**, 183–200.
Ulrich, R. K. (1970c). The five-minute oscillations on the solar surface. *Astrophysical Journal*, **162**, 993–1002.
Ulrich, R. K. (1976). A nonlocal mixing-length theory of convection for use in numerical calculations. *Astrophysical Journal*, **207**, 564–73.
Unno, W. (1969). Theoretical studies on stellar stability. II. Undisturbed convective non-grey atmospheres. *Publications of the Astronomical Society of Japan*, **21**, 240–62.
van der Borght, R. (1971). Finite amplitude convection in a compressible medium. *Publications of the Astronomical Society of Japan*, **23**, 539–51.
van der Borght, R. (1975a). Finite amplitude convection in a compressible layer with polytropic structure. *Australian Journal of Physics*, **28**, 437–52.
van der Borght, R. (1975b). Finite amplitude convection in a compressible medium and its application to solar granulation. *Monthly Notices of the Royal Astronomical Society*, **173**, 85–95.
van der Borght, R. & Waters, B. E. (1971). Non-linear convection with variable viscosity and thermal conductivity. *Proceedings of the Astronomical Society of Australia*, **2**, 47–8.
Wasiutynski, J. (1946). Studies in hydrodynamics and structure of stars and planets. *Astrophysica Norvegica*, **4**, 1–497.
Waters, B. E. (1971). Depth dependence of the various scale lengths in a quasi-Vitense model of the solar convection zone. *Proceedings of the Astronomical Society of Australia*, **2**, 48–50.
Weiss, N. O. (1977). Numerical methods in convection theory. In *Problems of Stellar Convection*, eds. E. A. Spiegel & J. P. Zahn, pp. 142–50. Berlin: Springer-Verlag.

Chapter 5

Altrock, R. C. (1976). The horizontal variation of temperature in the low solar photosphere. *Solar Physics*, **47**, 517–23.
Altrock, R. C. & Musman, S. (1976). Physical conditions in granulation. *Astrophysical Journal*, **203**, 533–40.

Ambrož, P. (1976). About the relation between the limb effect of the redshift on the Sun and the large-scale distribution of solar activity. In *Basic Mechanisms of Solar Activity* (IAU Symposium 71), ed. V. Bumba & J. Kleczek, pp. 113–18. Dordrecht: D. Reidel Publishing Co.

Beckers, J. M. & Nelson, G. D. (1978). Some comments on the limb shift of solar lines. II: The effect of granular motions. *Solar Physics*, **58**, 243–61.

Beckers, J. M. & Taylor, W. R. (1980). Some comments on the limb shift of solar lines. III. Variation of limb shift with solar latitude, across plages, and across supergranules. *Solar Physics*, **68**, 41–7.

Blackwell, D. E., Ibbetson, P. A., Petford, A. D. & Willis, R. B. (1976). Interpretation of the solar spectrum 300 nm to 900 nm – I. Fe I lines of excitation potential 0.00 eV–0.12 eV, microturbulence, damping, abundance, tests of oscillator strengths. *Monthly Notices of the Royal Astronomical Society*, **177**, 227–45.

Böhm, K. H. (1954). Zur Deutung der Mitte-Rand-Variation der Fraunhofer-Linien. *Zeitschrift für Astrophysik*, **35**, 179–202.

Brandt, P. N. & Schröter, E. H. (1982). On the centre-to-limb dependence of the asymmetry and wavelength shift of the solar line λ5576. *Solar Physics*, **79**, 3–18.

Bray, R. J. & Loughhead, R. E. (1967). *The Solar Granulation*. 1st edn. London: Chapman & Hall.

Bray, R. J. & Loughhead, R. E. (1978). On the possibility of measuring stellar convection parameters. *Publications of the Astronomical Society of the Pacific*, **90**, 609–14.

Bumba, V. & Kleczek, J. (eds.) (1976). *Basic Mechanisms of Solar Activity* (IAU Symposium 71). Dordrecht: D. Reidel Publishing Co.

Canfield, R. C. & Beckers, J. M. (1976). Observational evidence for unresolved motions in the solar atmosphere. In *Physique des mouvements dans les atmosphères stellaires*, ed. R. Cayrel & M. Steinberg, pp. 291–327. Paris: Centre National de la Recherche Scientifique.

Cloutman, L. D. (1979a). A physical model of the solar granulation. *Astrophysical Journal*, **227**, 614–28.

Cloutman, L. D. (1979b). The supergranulation: Solar rip currents? *Astronomy and Astrophysics*, **74**, L1–3.

Dodson, H. W. & Hedeman, E. R. (1968). Some patterns in the development of centers of activity, 1962–66. In *Structure and Development of Solar Active Regions* (IAU Symposium 35), ed. K. O. Kiepenheuer, pp. 56–61. Dordrecht: D. Reidel Publishing Co.

Dravins, D. (1974). Convection in the photosphere of Arcturus. *Astronomy and Astrophysics*, **36**, 143–5.

Dravins, D., Lindegren, L. & Nordlund, Å. (1981). Solar granulation: Influence of convection on spectral line asymmetries and wavelength shifts. *Astronomy and Astrophysics*, **96**, 345–64.

Durrant, C. J., Mattig, W., Nesis, A., Reiss, G. & Schmidt, W. (1979). Studies of granular velocities. VIII. The height dependence of the vertical granular velocity component. *Solar Physics*, **61**, 251–70.

Durrant, C. J. & Nesis, A. (1981). Vertical structure of the solar photosphere. *Astronomy and Astrophysics*, **95**, 221–8.

Durrant, C. J. & Nesis, A. (1982). Vertical structure of the solar photosphere. II. The small-scale velocity field. *Astronomy and Astrophysics*, **111**, 272–8.

Edmonds, F. N. (1962). A statistical photometric analysis of granulation across the solar disk. *Astrophysical Journal, Supplement Series*, **6**, 357–406.

Edmonds, F. N. (1964). Source function and temperature fluctuations in the solar photosphere. I. The isotropic approximation. *Astrophysical Journal*, **139**, 1358–73.

Edmonds, F. N. (1974). Convective flux in the solar photosphere as determined from fluctuations. *Solar Physics*, **38**, 33–41.

Gingerich, O. & de Jager, C. (1968). The Bilderberg model of the photosphere and low chromosphere. *Solar Physics*, **3**, 5–25.
Gingerich, O., Noyes, R. W., Kalkofen, W. & Cuny, Y. (1971). The Harvard-Smithsonian Reference Atmosphere. *Solar Physics*, **18**, 347–65.
Gray, D. F. (1980). Measurements of spectral line asymmetries for Arcturus and the Sun. *Astrophysical Journal*, **235**, 508–14.
Gray, D. F. (1981). Asymmetries in the spectral lines of Procyon. *Astrophysical Journal*, **251**, 583–4.
Gray, D. F. (1982). Observations of spectral line asymmetries and convective velocities in F, G, and K stars. *Astrophysical Journal*, **255**, 200–9.
Griffin, R. & Griffin, R. (1973). Accurate wavelengths of stellar and telluric absorption lines near λ7000 Å. *Monthly Notices of the Royal Astronomical Society*, **162**, 255–60.
Heintze, J. R. W., Hubenet, H. & de Jager, C. (1964). A reference model of the solar photosphere and low chromosphere. *Bulletin of the Astronomical Institutes of the Netherlands*, **17**, 442–5.
Howard, R. (1971). The large-scale velocity fields of the solar atmosphere. *Solar Physics*, **16**, 21–36.
Howard, R. & Bhatnagar, A. (1969). On the spectrum of granular and intergranular regions. *Solar Physics*, **10**, 245–53.
de Jager, C. (1954). High-energy microturbulence in the solar photosphere. *Nature*, **173**, 680–1.
de Jager, C. (1959). Structure and dynamics of the solar atmosphere. In *Handbuch der Physik*, vol. 52, ed. S. Flügge, pp. 80–362. Berlin: Springer-Verlag.
Jones, C. A. & Moore, D. R. (1979). The stability of axisymmetric convection. *Geophysical and Astrophysical Fluid Dynamics*, **11**, 245–70.
Kaisig, M. (1981). *Untersuchung von Asymmetrien photosphärischer Absorptionslinien in ruhigen und aktiven Gebieten der Sonne*. Diplomarbeit. Freiburg: University of Freiburg.
Keil, S. L. (1980a). The interpretation of solar line shift observations. *Astronomy and Astrophysics*, **82**, 144–51.
Keil, S. L. (1980b). The structure of solar granulation. I. Observations of the spatial and temporal behavior of vertical motions. *Astrophysical Journal*, **237**, 1024–34.
Keil, S. L. (1980c). The structure of solar granulation. II. Models of vertical motion. *Astrophysical Journal*, **237**, 1035–42.
Keil, S. L. & Canfield, R. C. (1978). The height variation of velocity and temperature fluctuations in the solar photosphere. *Astronomy and Astrophysics*, **70**, 169–79.
Kneer, F. J., Mattig, W., Nesis, A. & Werner, W. (1980). Coherence analysis of granular intensity. *Solar Physics*, **68**, 31–9.
Koch, A., Küveler, G. & Schröter, E. H. (1979). On depth-dependence of photospheric oscillations. *Solar Physics*, **64**, 13–25.
Margrave, T. E. & Swihart, T. L. (1969). Inhomogeneities in the solar photosphere. *Solar Physics*, **6**, 12–17.
Massaguer, J. M. & Zahn, J. P. (1980). Cellular convection in a stratified atmosphere. *Astronomy and Astrophysics*, **87**, 315–27.
Mihalas, B. W. & Toomre, J. (1981). Internal gravity waves in the solar atmosphere. I. Adiabatic waves in the chromosphere. *Astrophysical Journal*, **249**, 349–71.
Mihalas, B. W. & Toomre, J. (1982). Internal gravity waves in the solar atmosphere. II. Effects of radiative damping. *Astrophysical Journal*, **263**, 386–408.
Musman, S. & Nelson, G. D. (1976). The energy balance of granulation. *Astrophysical Journal*, **207**, 981–8.
Naze Tjötta, J. & Tjötta, S. (1974). Mass transport induced by time-dependent oscillations of finite amplitude in a stratified perfect gas. *Astronomy and Astrophysics*, **30**, 249–58.
Nelson, G. D. (1978). A two-dimensional solar model. *Solar Physics*, **60**, 5–18.

Nelson, G. D. (1980). Granulation in a main-sequence F-type star. *Astrophysical Journal*, **238**, 659–66.
Nelson, G. D. & Musman, S. (1977). A dynamical model of solar granulation. *Astrophysical Journal*, **214**, 912–16.
Nelson, G. D. & Musman, S. (1978). The scale of solar granulation. *Astrophysical Journal*, **222**, L69–L72.
Nordlund, Å. (1976). A two-component representation of stellar atmospheres with convection. *Astronomy and Astrophysics*, **50**, 23–39.
Nordlund, Å. (1977). Convective overshooting in the solar photosphere; a model granular velocity field. In *Problems of Stellar Convection*, eds. E. A. Spiegel & J. P. Zahn, pp. 237–8. Berlin: Springer-Verlag.
Nordlund, Å. (1978). Solar granulation and the nature of 'microturbulence'. In *Astronomical Papers dedicated to Bengt Strömgren*, eds. A. Reiz & T. Andersen, pp. 95–114. Copenhagen University Observatory.
Nordlund, Å. (1980). Numerical simulation of granular convection: effects on photospheric spectral line profiles. In *Stellar Turbulence*, eds. D. F. Gray & J. L. Linsky, pp. 213–24. Berlin: Springer-Verlag.
Nordlund, Å. (1982). Numerical simulations of the solar granulation: I. Basic equations and methods. *Astronomy and Astrophysics*, **107**, 1–10.
Pierce, A. K. & Breckinridge, J. B. (1973). The Kitt Peak table of photographic solar spectrum wavelengths. *Kitt Peak National Observatory Contribution*, No. 559.
Plaskett, H. H. (1955). Physical conditions in the solar photosphere. *Vistas in Astronomy*, **1**, 637–47.
Priestley, C. H. B. (1959). *Turbulent Transfer in the Lower Atmosphere*. University of Chicago Press.
Schröter, E. H. (1957). Zur Deutung der Rotverschiebung und der Mitte-Rand-Variation der Fraunhoferlinien bei Berücksichtigung der Temperaturschwankungen der Sonnenatmosphäre. *Zeitschrift für Astrophysik*, **41**, 141–81.
Simon, G. W. & Leighton, R. B. (1964). Velocity fields in the solar atmosphere. III. Large-scale motions, the chromospheric network and magnetic fields. *Astrophysical Journal*, **140**, 1120–47.
Simon, G. W. & Weiss, N. O. (1968). Supergranules and the hydrogen convection zone. *Zeitschrift für Astrophysik*, **69**, 435–50.
Sobolev, V. M. (1975). On the profile characteristics of the Fraunhofer lines in the granules and intergranular regions. *Solnechnye Dannye Bjulleten*, **6**, 68–71.
Spiegel, E. A. (1964). The solar hydrogen convection zone and its direct influence on the photosphere. In *Transactions of the International Astronomical Union*, vol. 12B, ed. J. C. Pecker, pp. 539–42. London: Academic Press.
Spruit, H. C. (1974). A model of the solar convection zone. *Solar Physics*, **34**, 277–90.
Turner, J. S. (1973). *Buoyancy Effects in Fluids*. Cambridge University Press.
Turon, P. (1973). Inhomogeneous model of the photosphere. *Solar Physics*, **41**, 271–88.
Voigt, H. H. (1956). Drei-Strom-Modell der Sonnenatmosphäre und Asymmetrie der Linien des infraroten Sauerstoff-Tripletts. *Zeitschrift für Astrophysik*, **40**, 157–90.
Voigt, H. H. (1959). Drei-Strom-Modell der Sonnenatmosphäre. II. Die infraroten Nickellinien λ7789 and 7798 Å. *Zeitschrift für Astrophysik*, **47**, 144–60.
Wilson, P. R. (1962). The application of the equation of transfer to the interpretation of solar granulation. *Monthly Notices of the Royal Astronomical Society*, **123**, 287–97.
Wilson, P. R. (1964). Photospheric structure and rms fluctuation data. *Astrophysical Journal*, **140**, 1148–59.
Wilson, P. R. (1969a). Temperature fluctuations in the solar photosphere. *Solar Physics*, **6**, 364–80.
Wilson, P. R. (1969b). Temperature fluctuations in the solar photosphere. II: the mean limb-darkening and the second maximum. *Solar Physics*, **9**, 303–14.

NAME INDEX

SUBJECT INDEX

(Note: unless otherwise apparent, entries refer to the *photospheric* granulation.)